Inconsistency, Asymmetry, and Non-Locality

OXFORD STUDIES IN THE PHILOSOPHY OF SCIENCE

General Editor
  Paul Humphreys, University of Virginia

Advisory Board
  Jeremy Butterfield
  Peter Galison
  Ian Hacking
  Philip Kitcher
  Richard Miller
  James Woodward

The Book of Evidence
  Peter Achinstein

Science, Truth, and Democracy
  Philip Kitcher

The Devil in the Details: Asymptotic Reasoning in Explanation, Reduction, and Emergence
  Robert W. Batterman

Science and Partial Truth: A Unitary Approach to Models and Scientific Reasoning
  Newton C.A. da Costa and Steven French

Inventing Temperature: Measurement and Scientific Progress
  Hasok Chang

Making Things Happen: A Theory of Causal Explanation
  James Woodward

Inconsistency, Asymmetry, and Non-Locality: A Philosophical Investigation of Classical Electrodynamics
  Mathias Frisch

INCONSISTENCY, ASYMMETRY, AND NON-LOCALITY

A Philosophical Investigation of Classical Electrodynamics

Mathias Frisch

2005

OXFORD
UNIVERSITY PRESS

Oxford University Press, Inc., publishes works that further
Oxford University's objective of excellence
in research, scholarship, and education.

Oxford   New York
Auckland   Cape Town   Dar es Salaam   Hong Kong   Karachi
Kuala Lumpur   Madrid   Melbourne   Mexico City   Nairobi
New Delhi   Shanghai   Taipei   Toronto

With offices in
Argentina   Austria   Brazil   Chile   Czech Republic   France   Greece
Guatemala   Hungary   Italy   Japan   Poland   Portugal   Singapore
South Korea   Switzerland   Thailand   Turkey   Ukraine   Vietnam

Copyright © 2005 by Oxford University Press, Inc.

Published by Oxford University Press, Inc.
198 Madison Avenue, New York, New York 10016
www.oup.com

Oxford is a registered trademark of Oxford University Press

All rights reserved. No part of this publication may be reproduced,
stored in a retrieval system, or transmitted, in any form or by any means,
electronic, mechanical, photocopying, recording, or otherwise,
without the prior permission of Oxford University Press.

Library of Congress Cataloging-in-Publication Data
Frisch, Mathias.
  Inconsistency, asymmetry, and non-locality: a philosophical investigation of classical electrodynamics / Mathias Frisch.
    p. cm.—(Oxford studies in the philosophy of science)
  Includes bibliographical references and index.
  ISBN-13   978-0-19-517215-7
  ISBN 0-19-517215-9
  1. Electrodynamics—Philosophy. I. Title. II. Series.

QC631.3.F75 2005
537.6'01—dc22   2004049286

9 8 7 6 5 4 3 2 1

Printed in the United States of America
on acid-free paper

For Andie and Julia

Preface

I began thinking about classical electrodynamics philosophically while I was working on my dissertation on scientific explanation and the role of models in scientific theorizing. I happened to come across a paper by Dudley Shapere (Shapere 1984b), in which he maintains that Lorentz's classical theory of the electron precludes the existence of point charges, while the theory of relativity requires that charged particles be pointlike. "Classical (relativistic) electrodynamics," Shapere concludes, "thus appears to contain a contradiction" (360). Yet despite this purported inconsistency, he holds that the theory is nevertheless useful; and this view seemed to fit well with the thesis I tried to defend in the dissertation: that a theoretical scheme need not be true in order to be genuinely explanatory. I have spent the last few years trying to understand classical electrodynamics better and, among other things, have tried to locate the inconsistency more precisely. In the end, I think Shapere should find my conclusions sympathetic to his own views, even though some of the details of my investigation might not quite support his evocative if rather sketchy remarks on the theory. The seeds for this investigation, then, were planted during my work on my dissertation, and I would like to thank my advisers at Berkeley, Elisabeth Lloyd and Martin Jones, for their support, guidance, and friendship.

Prior to reading Shapere's paper I had had ample exposure to classical electrodynamics as a physics student. I had taken five semesters of electrodynamics at the advanced undergraduate and graduate level, at the University of Munich and the University of California at Berkeley, without ever noticing that the theory might in any way be conceptually problematic. This might simply have been due to my own lack of attention. But I would like to believe that there is a more interesting and more general explanation for this—that one can learn quite a bit of physics and can acquire the skills to successfully apply a theory to a large number of phenomena in its domain without ever worrying about the overall consistency of the approach one is using. Certainly, the conceptual problems of the theory were not emphasized in these courses, and it is hard to detect them in standard electrodynamics textbooks. If the problems are mentioned at all, it is as an aside that in no way interferes with the theory's usefulness, and thus can be ignored by physics students—who, after all, need to think mainly about solving problem sets.

The three science texts from which I learned most about the structure and content of classical electrodynamics are John Jackson's *Classical Electrodynamics* (Jackson 1975), the "Bible" of classical electrodynamics, which is now in its third edition (Jackson 1999); Fritz Rohrlich's *Classical Charged Particles* (Rohrlich 1965; 1990); and Steven Parrott's *Relativistic Electrodynamics and Differential Geometry* (Parrott 1987). (The last of the three is written by a mathematician who focuses explicitly on the conceptual problems of the theory.) As far as the problem of the asymmetry of radiation is concerned, I found H. Dieter Zeh's *The Physical Basis for the Direction of Time* (Zeh 2001) the most useful. I am extremely grateful to all four authors for very helpful E-mail exchanges over the last few years.

Most of the work for the book was undertaken while I was an assistant professor at Northwestern University. I would like to thank Arthur Fine and David Hull for many helpful conversations and for their support, encouragement, and advice. For two of my years at Northwestern, I had the good fortune to have as a colleague Eric Winsberg, who was a postdoctoral fellow in the Program in History and Philosophy of Science; I want to thank him for many, many hours of spirited debates and conversations. In general, Northwestern's HPS program offered a wonderful and intellectually stimulating environment for doing work in philosophy of science, and it is a shame that the program has been discontinued.

Large portions of this book were written while I was on leave from Northwestern with fellowships from the National Science Foundation (award no. SES-0093212) and the National Endowment for the Humanities (award no. FA-36796-01). I am extremely grateful to both agencies for their generous support.

My leave time was spent in southern California, and I would like to thank Jim Woodward, Alan Hajek, and Chris Hitchcock at Cal Tech, and Jeff Barrett and David Malament at UC Irvine, for their hospitality and helpful conversations during my time there. I am especially thankful to David Malament for probing criticisms of my central claims in chapters 2 and 3. I am afraid that he has still not convinced me, but I hope my arguments have gotten stronger as a result of our discussions.

I first got interested in the problem of the arrow of radiation when I read Huw Price's book *Time's Arrow and Archimedes' Point* (Price 1996), and subsequently discussed and criticized his account of the asymmetry (Frisch 2000). My thinking on this issue has significantly evolved since then (I hope); and to the extent that it has, this is largely due to discussions I have had with Huw (despite the fact that we still do not fully agree on all issues). I am especially grateful for his invitation to me to visit the Centre for Time at the University of Sydney for a short stay in the fall of 2002. I want to thank both Huw and the physicist David Atkinson for the many stimulating and fruitful conversations there (and for introducing me to Sydney's excellent food and coffee!).

In addition to those mentioned above, I would like to thank Marc Lange for detailed comments on earlier drafts of several chapters, and Adolf Grünbaum for critical comments on a draft of chapter 4. Finally, through the years Nancy Cartwright and Paul Teller have always been very encouraging and supportive of this work. I have learned a lot both through studying their writings and through conversations and discussions with them. I owe both of them a tremendous amount.

Contents

Preface   *vii*

    Chapter 1: Introduction: Theories and Models   3

Part One: Particles   23
    Chapter 2: Inconsistency   25
    Chapter 3: In Search of Coherence   47
    Chapter 4: Non-Locality   73

Part Two: Fields   101
    Chapter 5: The Arrow of Radiation   103
    Chapter 6: Absorber and Entropy Theories of Radiation   121
    Chapter 7: The Retardation Condition   145
    Chapter 8: David Lewis on Waves and Counterfactuals   165

    Chapter 9: Conclusion   193

Notes   *195*
Bibliography   *203*
Index   *209*

Inconsistency, Asymmetry, and Non-Locality

Introduction: Theories and Models

1. Introduction

I have two main aims in this book. The first is within the philosophy of physics: I want to investigate certain aspects of the conceptual structure of classical electrodynamics. This theory has been largely ignored by philosophers of science, probably at least partly due to the mistaken view that it is conceptually unproblematic. For much of the history of the philosophy of physics, philosophers have been interested mainly in quantum physics and, to a somewhat lesser extent, the theory of relativity. While there are good reasons for this narrow focus, it has come at a cost in that it has led to somewhat of a caricature view of classical physics, part of which is the belief that classical physics is philosophically uninteresting. Thus, when philosophers mention classical electrodynamics at all, it usually is as the paradigm of a causal and deterministic classical theory. As I will argue, this perception of the theory as satisfying all the conditions on the methodologist's wish list, as it were, is wide of the mark. In fact, the most common theoretical approach to modeling the interactions between charged particles and electromagnetic fields is mathematically inconsistent despite the fact that it is strikingly successful.

Its conceptual problems notwithstanding, the classical theory of particles and fields is one of the core theories of modern physics, and it provides a fruitful case study for investigating a number of general methodological and metaphysical issues in the philosophy of science. My second aim, then, is to try to show that investigating a particular scientific theory in some detail can shed light on concerns in the general philosophy of science. In particular, I want to appeal to certain aspects of classical electrodynamics to challenge what still appears to be the standard conception of scientific theories among philosophers of physics as well as philosophers of science more generally.

This standard conception consists of a cluster of loosely connected views. Briefly put, it holds that a scientific theory (at least in the physical sciences) consists of a mathematical formalism and an interpretation. At the core of the formalism are the theory's laws that define the class of models of the theory. The job of the interpretation is to specify a mapping function from bits of the mathematical formalism to bits of the world, and thereby to map the models of the theory onto the possible worlds allowed by the theory. That is, on the standard conception, the laws of the theory—understood as interpreted expressions—delineate the range of what, according to the theory, is physically possible. Importantly, on the standard conception, fixing the ontology of a theory exhausts the role of an interpretation.

I have two main complaints about this conception. First, the condition of consistency is built into this account of theories (as we will see in more detail below). One of the theses of this book is that consistency is just one criterion of theory assessment among several, and that a theory or theoretical scheme can be successful in representing the phenomena within its domain of application without being consistent. For inconsistent 'laws' to be applicable empirically, building models of the phenomena cannot proceed in the manner suggested by the standard conception—that is, merely by plugging an appropriate set of initial and boundary conditions into the theory's laws. Rather, in addition to the laws themselves, a theory has to provide us with (perhaps implicit) rules on how to apply these laws. The standard conception, then, is too 'thin,' in that there are important aspects of scientific theorizing that are left out by this account.

Second, the conception is too thin in another respect: It is broadly neo-Humean or neo-Russellian in that it does not allow for rich or 'thick' causal notions that cannot be spelled out in terms of the mathematical formalism plus a mapping function to be part of the content of scientific theories. Neo-Russellians hold that a fundamental physics with time-symmetric laws only supports a notion of time-symmetric functional determination of states at one time by states at other times, but does not support an asymmetric notion of causes 'bringing about' their effects. I will argue that this view is mistaken and that thick asymmetric causal concepts can play an important role even in fundamental physics. My argument is developed mainly through a detailed discussion of the *problem of the arrow of radiation*. Radiation phenomena exhibit a temporal asymmetry. It will require some care to spell out exactly what this asymmetry consists in, but roughly and in a preliminary way it can be expressed as follows: There are coherent, diverging electromagnetic waves in nature, but not coherent, converging waves, and this is so despite the fact that the wave equation that can be derived from the Maxwell equations—the equations at the heart of classical electrodynamics—allows for both types of waves.

After providing a comprehensive survey and a critique of the major attempts by both physicists and philosophers to solve the puzzle, I will argue that the asymmetry is best thought of as a causal asymmetry. Thus, the theory essentially contains certain causal assumptions which go beyond what is embodied in the theory's mathematical formalism and, hence, any attempt at a neo-Humean or neo-Russellian construal of classical electrodynamics must fail. Additional evidence for the impossibility to excise causal notions from fundamental physics, I will argue, is provided by the fact that physicists employ various irreducibly causal locality conditions in characterizing the content of physical theories. Thus, as in the case of the inconsistency, my conclusion is that it is a mistake to identify the content of a scientific theory with its mathematical formalism plus a mapping function. A theory's interpretive framework can be richer than the standard account allows.

In this introductory chapter I will provide a brief survey of philosophical accounts of scientific theorizing, focusing on the role of consistency conditions in these accounts. In the last section of the chapter I will give a preview of the claims and arguments of subsequent chapters.

2. From Coherent Worldviews to Maps

The twentieth century produced two major types of philosophical accounts of scientific theories—*sentence views*, on the one hand, and various versions of a *semantic view*, on the other. Both types provide identity conditions for theories intended as philosophical reconstructions of the concept of scientific theory. That is, they define a philosophical concept that (in the spirit of the project of philosophical analysis) is meant to allow us to capture what is philosophically interesting about scientific theorizing. As we will see, despite the apparent differences between the two types of views, both agree in their commitment to what I have called 'the standard account.'

According to sentence views, scientific theories are to be identified with deductively closed sets of sentences. On an early version of the view, these sentences are to be formulated in a first-order formal language and are given a partial interpretation via a set of *correspondence rules*. On a later version, theories are taken to be formulated in a natural language that is assumed to be antecedently understood.[1]

Against the syntactic view, the more recent semantic view of theories maintains that scientific theories ought to be identified with certain abstract nonlinguistic structures. Advocates of the semantic view urge us to shift our attention from the linguistic formulation of a theory (and the problems that attempts at reconstructing scientific theories in a first-order formal language have engendered) to nonlinguistic structures that are picked out by the sentences of a theory. Focusing directly on these structures, advocates of the semantic view maintain, will lead to a philosophical reconstruction of scientific theories that does better justice to scientific practice than does the syntactic view. In particular, giving up attempts at formalizing theories syntactically allows philosophers, in characterizing the structures constituting a theory, to use the same language that scientists themselves are using—that is, in the case of physics, the language of mathematical physics. These abstract structures are characterized differently in various formulations of the semantic view, but all proponents of the semantic view agree that the structures should be thought of as models in some sense. Thus, Bas van Fraassen says: "To present a theory is to specify a family of structures, its *models*" (van Fraassen 1980, 64; italics in original), while Frederick Suppe maintains that "theories are extralinguistic entities" such that a "theory qualifies as a model for each of its formulations" (Suppe 1977, 222). And Ronald Giere proposes that "we understand a theory as comprising two elements: (1) a population of models, and (2) various hypotheses linking those models with systems in the real world" (Giere 1988, 85).[2]

Central to the semantic view is the notion of a model. Yet one source of confusion has been (and continues to be) that the early literature on the view conflated (at least) two distinct notions of model, with the unfortunate consequence that much of the usefulness of introducing talk of models into philosophical discussions of scientific theorizing was lost. On the one hand, advocates of the semantic view appeal to a notion of model derived from logic and model-theory, where a model of a set of sentences is a structure in which the sentences are true. Van Fraassen, for example, defines the notion of model as follows: "Any structure which satisfies the axioms of a theory...is called a *model* of that theory" (van Fraassen 1980, 43).

On the other hand, the semantic view takes models to be *representations* of phenomena. For example, the Bohr model of the atom (which van Fraassen also discusses) is a model in virtue of the fact that it is intended to represent atoms and certain aspects of their behavior. Here it is immaterial whether we identify the Bohr model with a set of equations or with some abstract, idealized structure picked out by these equations. Whatever thing the Bohr model is, it is a model in virtue of its representational function. (It is also worth pointing out that the notion of representation here cannot be reduced to that of resemblance. We may plausibly hold that the empirical success of a model depends on whether it at least partially resembles the thing it represents in certain relevant respects and to a certain degree. But quite obviously, I think, resemblance is neither necessary nor sufficient for representation.) Important for our purposes here is that the notions of model inspired by model-theory and the notion of model as representation are quite distinct. According to the first notion, an entity is a model in virtue of its relation to a set of sentences, while on the second notion, an entity is a model in virtue of its relation to the world.[3] If one is not careful in drawing this distinction, it will probably strike one as somewhat mysterious how an inconsistent theory which has no *model-theoretic* models can nevertheless provide us with *representational* models of the phenomena.

What are these structures which the semantic view sees at the heart of scientific theories? Following van Fraassen, it has become customary to take a theory's models to be *state space models*. State space models are 'model universes' that a theory's mathematical formalism allows us to construct and that represent the phenomena within the theory's domain. For practical reasons we are usually interested only in very simple universes that have only as much structure as is needed to represent a particular phenomenon individually, such as an isolated planetary system, a pendulum, or a single charged particle in an external electromagnetic field. The model universes specify the possible states of a system, where the state is given by an assignment of values to a set of dynamical variables. In the case of a mechanical universe consisting of n particles, one possible choice for these variables would be the $3n$ Cartesian position coordinates together with the corresponding $3n$ momentum coordinates. The space spanned by the $6n$ variables is the system's *state space* (or *phase space*). In the case of a universe containing continuous fields, the state involves an infinity of variables whose possible values represent the state of the field at all points in space. Given the state at a particular time, the equations of motion for a system, as given by the mathematical formalism, determine a possible temporal evolution for the system. The equations of motion delineate possible histories of the model universe, where different particular histories are picked out by different *initial conditions*—that is, different states at a particular time. A particular trajectory in state space represents a possible history of the system. A theory is deterministic if it is the case that if the states of two systems agree at one time, they agree for all times. Van Fraassen maintains that we ought to strictly identify a theory with the class of its state space models. Thus, van Fraassen's view is a version of the standard account, according to which the content of a theory is exhausted by its mathematical formalism and a mapping function defining the class of its models.

In fact, versions of the semantic view that emphasize a notion of model analogous to that in model-theory have much in common with syntactic views.[4] In particular, on both types of view, consistency comes out as a crucial condition for scientific theories. If theories are identified with deductively closed sets of sentences, then consistency is a necessary condition for a set of sentences being an even minimally successful theory (if we assume that inferences from the theory are governed by classical logic). By the same token, if we think of a theory's models as structures in which the theory's laws or axioms are true, then the laws of the theory need to be consistent. For a theory with inconsistent laws has no models.

According to the two major philosophical accounts of scientific theories, internal consistency is built into the definition of 'scientific theory' and comes out as a necessary condition of something's being a scientific theory. Why is this so? Why do both types of account preclude from the outset the possibility of a genuine theory that is inconsistent? It appears that the importance of consistency is supported by considerations that are far more widely accepted than any particular philosophical analysis of the concept of theory. The aim of a scientific theory, it is commonly held, is to present us with a coherent account of what is physically possible. Accordingly, a theory ought to present the phenomena in its domain as being governed by a set of principles, the theory's laws, such that representations of different phenomena can be integrated into a unified and coherent account of ways the world could be.

While in practice we may often be interested only in how a given theory represents certain phenomena individually, it should always be possible—at least in principle—to integrate representations of different phenomena into coherent, physically possible worlds. That is, the view says that theories do not merely provide us with state space models of individual phenomena, but that for each set of phenomena in the theory's domain there will always be a larger state space in which the phenomena in question can be given a unified representation (as long as there are possible initial conditions that are compatible). For example, since Newton's theory defines state spaces for representing the behavior of pendulums and also state spaces for representing planetary systems, then the theory should also (in principle) allow us to construct larger state space models for a pendulum on the surface of a planet revolving around a star. This larger model universe will again be a Newtonian system. Possible histories of the pendulum subsystem within the larger system will not strictly agree with those of an isolated pendulum in a constant gravitational field, but the disagreement will be practically insignificant. In fact, it is this last fact that licenses our treatment of a pendulum as an isolated system: For most purposes, variations in the gravitational force acting on the pendulum can be ignored.

As far as this view is concerned, it does not matter whether we think of theories syntactically or semantically. That is, it does not matter whether we choose to identify a theory with a mathematical formalism which serves to define a class of state space models, or whether we identify the theory with the class of models. Yet today the view is most often expressed by appealing to the semantic notion of possible worlds. To quote van Fraassen again, we "can think of [a theory's state space] models as representing the possible worlds allowed by the theory; one of the possible worlds is meant to be the real one" (van Fraassen 1980, 47).

While one might be tempted to think that the talk of possible worlds here is meant only metaphorically and that the worlds allowed by a theory need not be anything more than such simple 'worlds' as isolated pendulums or pairs of colliding balls, van Fraassen says quite explicitly that the possible worlds delineated by a theory are not merely the impoverished model universes representing particular phenomena. Rather, the models with which a theory provides us include models rich enough to represent possible worlds as complex as ours. And I think this assumption is implicit in any account of theories in terms of possible worlds, such as that of John Earman, who invokes the framework of possible worlds in his examination of determinism (Earman 1986) by asking whether certain "classical worlds" and "relativistic worlds" are deterministic. Similarly, Gordon Belot has distinguished three components of a physical theory—formalism, interpretation, and application—which he characterizes as follows:

> The formalism is some (more or less rigorous) mathematics.... The application is a set of practices which allow one to derive and to test the empirical consequences of the theory. The interpretation consists of a set of stipulations which pick out a putative ontology for the possible worlds correctly described by the theory. (Belot 1998, 533)[5]

Why is it not enough for a theory to provide us with a 'grab bag' of perhaps loosely connected models that resist being integrated into models of complex possible worlds? Why should we think that theories ought in principle to provide us with models of the world as a whole? One possible motivation for this view seems to be the idea that the ultimate aim of science, or at least of physics, is to find theories that are candidates for a true theory of everything. Theory change in physics, then, might appear to be driven by the pursuit of this aim: Theories are proposed and adopted as possible candidates for a correct universal physics until they are replaced by more promising successors. Theory change, on this picture, is a change of worldview. For example, the long-reigning mechanical worldview of Newtonian physics was replaced at the end of the nineteenth century by a (classical) electromagnetic world picture, which in turn was overthrown by a quantum mechanical view of the world. But to be a candidate for a true universal physics, a theory minimally has to present us with a coherent account of what the world might be like. And this requires that a theory's mathematical formalism be consistent. Thus, consistency, on this picture, turns out to be a minimal condition a theory must satisfy to be successful.

But there is a much more mundane reasons why one might think that scientific theories have to be consistent. If drawing inferences from a theory in science is anything like deductive inference in logic, then an inconsistent theory is trivial in the sense that any arbitrary sentence can be derived from it. It is a necessary condition for a theory to have content with which the theory's mathematical formalism, with the theory's laws at its core, is consistent. For if the laws of the theory are inconsistent, then there are no possible worlds in which they could hold simultaneously and the class of possible worlds allowed by the theory is empty. Since, according to the standard conception, the content of a theory is given by the possible worlds correctly described by the theory, the view implies that an inconsistent theory can have no content. Conversely, this means that if an inconsistent theory is to have a nontrivial interpretation, then that interpretation cannot consist in a description of the possible worlds in which the theory is true.

Thus, among the many different criteria scientists use to evaluate and compare theories, internal consistency seems to be privileged. One famous, standard list of such criteria is that proposed by Thomas Kuhn (1977, 321–322), who identifies accuracy, consistency, scope, simplicity, and fruitfulness as criteria for evaluating the adequacy of a theory.[6] Kuhn's criterion of consistency has two aspects—internal consistency and consistency with the theoretical context or framework within which a theory is proposed. The context is given both by other accepted theories and by general '*meta*physical,' methodological, or conceptual demands (such as those embodied in certain causality or locality conditions, I want to maintain). According to Kuhn, theories exhibit these characteristics to varying degrees, and the theory is most successful that scores highest on some weighted average of the criteria. He insisted, however, that there is neither a general recipe for applying each criterion individually nor a universal procedure for determining the relative importance of the different criteria.

Now, in computing the average 'score' of a theory, internal consistency seems to play a special role (even though Kuhn himself might not have thought so), for it seems that, according to the views discussed above, a theory cannot possibly be successful unless it is consistent. For internal inconsistency seems to negatively affect how high the theory scores on several of the other criteria, such as empirical accuracy and consistency with other theories. An inconsistent theory cannot be consistent with any other theory, and if any sentence whatever can be derived from such a theory, the theory will score very low on the condition of accuracy (since it appears to allow for arbitrarily inaccurate predictions to be derived from it). Thus, while even our most successful theories may exhibit any of the other criteria to varying degrees—a lack of simplicity, for example, might be made up for by gains in accuracy, or vice versa—internal consistency seems to be a necessary condition for a theory to be even minimally successful.

I want to argue that this view is mistaken. Consistency is just one criterion of theory assessment among many, and just as in the case of each of the other criteria, a lack of consistency can be made up for by a higher score on the remaining criteria. A theory or theoretical scheme, that is, can be successful, and even be the most successful scheme in its domain, without being consistent.

Now, there is one relatively recent view of scientific theories that can readily accommodate inconsistent theories. This view argues that a theory's laws do not 'hook up' to the world directly, as it were, but that the relation between laws and the world is 'mediated' by one or multiple layers of representational models of the phenomena. Due to its emphasis on the role of models in scientific theories, this view can be thought of as a variant of the semantic view, but it disagrees with other versions of that view in crucial respects. As we have seen, according to a view like van Fraassen's, the structures that theories postulate to represent particular phenomena do double duty: They are representational models of the phenomena and at the same time are model-theoretic models of the laws of the theory—that is, the theory's laws are true in those structures. By contrast, advocates of the rival 'model-based' view, such as Nancy Cartwright (1983, 1999; Cartwright, Shomar, and Suarez 1995), Ronald Giere (1988, 1999a, 1999b), and Margaret Morrison (1999),[7] argue that 'low-level' representational models of particular phenomena are in general not structures in which the theory's laws are true and, thus, that such models have an existence that is in some sense independent of the theory's laws.[8]

While earlier views emphasized globalizing or universalizing aspects of scientific theorizing, the recent model-based accounts focus on the role of scientific theories in constructing 'localized' models of particular phenomena. Instead of demanding that a theory define possible worlds as rich as ours, these accounts direct our attention to the practices that go into constructing models of single, isolated phenomena and argue that this practice suggests that many of even our best theories might well not provide us with any more comprehensive (and empirically well-supported) models of what the world as a whole may be like. In other words, rather than construing theories as providing us with 'global' accounts of ways the world could be, theories are understood as aiding us in the construction of 'local' models.

Advocates of the newer model-based view insist that a theory's laws do not determine the models that scientists use to represent the phenomena. According to the standard conception, scientific theories need to consist of no more than a mathematical formalism and a mapping function to successfully represent the world. Cartwright calls this view the "vending machine view," according to which a theory's mathematical formalism simply spits out a fully formed representation as the result of feeding in the appropriate input (Cartwright 1999, 184). The input consists of sets of initial or boundary conditions, fed into the theory's fundamental laws (which take the form of differential equations), and the output consists of a possible history for the system in question. On the standard conception, the only obstacle to generating models for every phenomenon of interest is that the system may be too large, and the initial and boundary conditions too complex, to result in a manageable system of equations. Cartwright contrasts the "vending machine view" with her own view that in many interesting applications of physics, scientists construct models "that go beyond the principles of any of the theories involved" (Cartwright 1999, 182). Building testable models, according to the model-based view of theories, usually involves highly context-dependent idealizing and approximating assumptions, and often requires appealing to assumptions from a number of different and sometimes incompatible theories. Since models are constructed with the help of a variety of resources not confined to any one theory, where the job of the scientist is to choose the items appropriate for the particular task, "models, by virtue of their construction, embody an element of independence from both theory and data" (Morrison and Morgan 1999, 14).

Since high-level laws can generally be obtained from observable phenomena only with the help of abstractions and idealizations, these laws, strictly speaking, 'lie' about the phenomena they are meant to govern, if they are construed as making claims about the world (see Cartwright 1983). Alternatively, advocates of a model-based view have also suggested that the laws of a theory should not be construed as having representational character at all. Instead, we ought to think of them as "tools" for model-building (Cartwright, Shomar, and Suarez 1995) or as "recipes" for constructing models (Giere 1999a). On either variant, theories do not have a tight deductive structure, and the 'models of a theory' need not be related to each other by more than "family resemblances" (Giere 1988).

Giere likens scientific representations to maps. Maps represent their objects only partially: "There is no such thing as a complete map" (Giere 1999a, 81; see also 25). Different maps present us with different perspectives on the world, embodying different interests. Maps correspond in various ways to the world and are more or less accurate, but no map represents all features of a region of the world and no map is perfectly accurate. Moreover, the cost of representing certain features either more accurately or in more useful ways (to a given purpose) often is that other features are represented less accurately. All these features, Giere maintains, representational models share with maps. The metaphor of scientific representations as maps strongly suggests the view that these representations need not be mutually consistent: Different maps have different purposes with their own standards of adequacy and, one may think, there need not be a guarantee that different maps that are each individually successful (in light if their different purposes) in representing a certain region of the world are mutually 'consistent' or 'congruent' with each other.

Now one might maintain that the disagreement between recent model-based views and the standard conception is to a large extent only a disagreement of emphasis. Traditionally, philosophers of science have not been much interested in theory application. The recent model-based view argues that ignoring experimental applications of theories is a mistake and that examining the interaction between theory and experiment is a philosophically fruitful enterprise. Yet one can accept this claim as well as the more specific claims of the model-based view concerning modeling practices in the sciences, and at the same time continue to believe that the content of a theory is given by the class of state space models defined by the theory's laws. That is, when advocates of the standard conception, on the one hand, and proponents of the model-based view, on the other, are making seemingly contradictory claims about models that we can construct with the aid of a theory, they may in fact be talking about two different things: here, about the pure, 'rarefied' state space models of the theoretician, and there, about the messy representational structures scientists devise with much ingenuity and creativity to enable their theories to make contact with the phenomena.

Genuine disagreement exists, however, at least over the issue of theory acceptance. According to Cartwright, the fact that modeling requires a host of highly context-specific assumptions and that empirically testable models can be constructed only for very special circumstances, suggests that we have no warrant for extending our commitment even to our best theories beyond the isolated pockets within which they have been used successfully in model-building. In light of the severely restricted domains of applicability of theories, the project of trying to arrive at a universal physics seems to Cartwright to be misguided.[9] By contrast, advocates of the standard conception generally appear to believe that the support our theories acquire within the confines of the laboratory also extends to the world at large.

I have said that a certain amount of peaceful coexistence between the two rival views of theories is possible, if we realize that they might be talking about different aspects of scientific theorizing. Yet the case of inconsistent theories suggests that certain features of scientific theorizing to which the model-based view points make their appearance even at the very top, as it were—at the level of pure theory, far removed from any concerns with actual experimental application. For if there are genuine theories with inconsistent laws, then, as we will see in more detail in the next chapter, the laws of the theory cannot on their own determine the theory's models, and there need to be constraints in addition to the laws that govern the laws' application. The theory's laws do not delineate a coherent class of possible worlds in this case, and the theory has no model-theoretic models that could double as representational models. Inconsistent theories, then, may be taken to provide particularly strong support for the importance of 'model-based' accounts of theories. We will also see (in chapter 3) that the kind of pragmatic considerations one might associate more commonly with 'model-building' make their appearance in classical electrodynamics even at the level of trying to arrive at a fundamental equation of motion. Thus, I take my discussion in the following chapters to be broadly supportive of a 'model-based' account of theories. If we were to identify theories with an interpreted formalism, then Cartwright's claim that constructing models often goes "beyond the principles of any of the theories involved" is supported even by 'high-level' modeling within classical electrodynamics.

I want to stress, however, that I do not wish to argue here for any of the general metaphysical conclusions Cartwright and others might wish to draw from their account of theorizing—such as that the world is truly "dappled" (Cartwright 1999)—and I also do not want to claim that there is a single account of theories that can adequately capture all aspects of scientific theorizing. I do not wish to deny the usefulness of formalizations of theories that do, where possible, investigate the possible worlds allowed by a theory. Sometimes scientists do seem to be interested in global representations of ways the world might be, and in such a case a possible worlds account of theories may well be philosophically illuminating. My complaint here is directed only against the view that such formalizations can provide us with identity conditions of what, in the spirit of the project of conceptual analysis, a scientific theory is.

It is sometimes argued that inconsistent theories may play a useful preliminary role in theory development, but that our best theories 'at the end of inquiry' could not be inconsistent. For example, Philip Kitcher has argued in an exchange with Helen Longino that even though model-based accounts of science are correct in maintaining that science presents us with "a patchwork of locally unified pieces," the suggestion that these pieces may be mutually inconsistent is problematic

(Kitcher 2002a, 2002b).[10] According to Kitcher, giving up the idea of science striving toward a single correct, all-encompassing theory of the world and allowing for multiple systems of representation geared toward different interests and perspectives does not force us to accept the possibility of inconsistent representations. Indeed, he suggests that it is the very fact that our representations are partial or incomplete that may enable different representations reflecting different interests to be nevertheless mutually consistent. Kitcher acknowledges that there are examples in the history of science of representations, accepted at the same time, that are not consistent. Yet he urges us to draw a distinction between the representations accepted by scientists at any stage in the history of science and the representations that "conform to nature," such as "the true statements, the accurate maps, the models that fit parts of the world in various respects to various degrees" (Kitcher 2002a, 570–571). That the former may be inconsistent is a commonplace, Kitcher thinks, but the latter, he claims, are jointly consistent.[11]

How are we to decide whether there are inconsistencies only between representations that are provisionally accepted at a given time or whether even partial models that "conform to nature" can be mutually inconsistent? For any example of actual inconsistent representations, it always appears to be open to the methodological conservative to maintain that these representations are only provisional and that we have not yet found those representations which do conform to nature. What, then, could possibly constitute a counterexample to Kitcher's claim that the representations that conform to nature are jointly consistent? One question that arises at this point is which side in the debate has the burden of proof. On the one hand, methodological radicals might try to argue that the history of science is full of examples of mutually inconsistent successful representations (even though much work would need to be done to establish this claim convincingly) and that, thus, Kitcher's belief in the possibility of jointly consistent representations is simply a declaration of faith. In fact, this is what Longino maintains (Longino 2002c, 563). On this view, it is up to the conservative to provide a convincing argument for the claim that our best representations will be consistent. On the other hand, Kitcher's tenet that representations conforming to nature are jointly consistent might appear simply to be a consequence of the claim that 'nature itself could not be inconsistent,' and the idea that inconsistent representations could all truthfully represent nature may be an idea that even many methodological radicals are unwilling to accept.

There is, however, an ambiguity in Kitcher's notion of a representation's conforming to nature. On the one hand, Kitcher cites true statements as examples of such representations, and quite plausibly sets of true statements would have to be jointly consistent. Thus, if the aim of science was to discover true statements, then he would be correct in maintaining that such representations could not be inconsistent with each other. Yet on the other hand, Kitcher also takes "models that fit parts of the world in various respects to various degrees" to be the kind of thing that can conform to nature. But once we allow that even models which fit the world only to various degrees can in the relevant sense conform to nature, conforming yet inconsistent representations become possible. Nature itself may be consistent, as it were, even though representations conforming to nature need not be jointly consistent. In particular, the models that may be most successful at balancing various conditions of adequacy need not be those which fit the world most accurately.

What is essentially the same point can also be put in the following way. The fact that Kitcher takes maps and 'imperfect' models to be examples of representations that can conform to nature suggests that for him, conformance does not require perfect fit. But if what Kitcher meant by a representation's conforming to nature implies that the representation fits perfectly, then his distinction between in some sense provisional and possibly inconsistent representations, on the one hand, and representations that conform to nature, on the other, is not exhaustive. For what this distinction does not recognize is the possibility of mutually inconsistent representations that are *not* provisional but represent an end product of scientific theorizing in that they provide us with the best possible balance of various conditions of adequacy or success—a balance which is unlikely to be improved upon. That is, what Kitcher's distinction does not allow for is the possibility that the representations of the phenomena in a certain domain that are most successful in balancing various theoretical virtues, such as those proposed by Kuhn, are mutually inconsistent, and yet it is unlikely that further research will lead to both more successful and jointly consistent representations.

So far I have focused on the role of a consistency condition in accounts of scientific theorizing. I want to end this section with some much briefer remarks on the notion of causation. If, with van Fraassen and others, we identify a theory with the class of its state space models, then causal assumptions cannot be part of the content of physical theories with time-symmetric laws. Van Fraassen is explicit about this point. In response to the question raised by Cartwright, "Why not allow causings in the models?," he replies:

> To me the question is moot. The reason is that, as far as I can see, the models which scientists offer us contain no structure which we can describe as putatively representing causing, or as distinguishing between causings and similar events which are not causings.... The question will still be moot if the causes/non-causes distinction is not recoverable from the model. Some models of group theory contain parts representing shovings of kid brothers by big sisters, but group theory does not provide the wherewithal to distinguish those from shovings of big sisters by kid brothers. The distinction is made outside the theory. (van Fraassen 1993, 437–438)

The distinction between causes and non-causes, van Fraassen goes on to say, can be drawn only "extra-scientifically." I have quoted this passage at length because it is a clear and explicit expression of the view against which I will argue. Theories, for van Fraassen, are identified with their models, and any claims that are not recoverable from these models cannot be scientific.

Why might one want to exclude causal assumptions from the content of a theory from the outset? One might want to reply simply by reiterating that causal assumptions are not part of the models. Yet this answer sidesteps the more basic question of why we should strictly identify theories with the class of their models and take the distinction between what is recoverable from the models and what is not, to delimit what is scientific from what is not. One motivation might be that we have strong faith in the tight match between our mathematical tools and the world, expressed in Galileo's dictum that the language in which the book of nature is written is mathematics. If the mathematical formalism of our best theories does not provide us with the means for representing causal relationships, then perhaps causal relations cannot be part of the world.

But we will see in our investigation of the inconsistency of classical electrodynamics that there are independent reasons for doubting that there must be such a tight fit, as it were, between our best mathematical formalism and the world. Moreover, as Judea Pearl has argued, even if systems of equations alone are insufficient to capture causal relationships, it may well be possible to express causal relations formally through equations in conjunction with directed graphs (Pearl 2000). If Pearl is right, this presents further evidence that we ought to be cautious in trying to draw inferences from the limits of any one particular mathematical formalism to limits of what is real. That group theory does not allow us to represent causings may point to a limit of group theory but does not yet show that the distinction between causes and non-causes is scientifically illegitimate.

Another motivation for excluding anything that is not recoverable from the models seems to be empiricist scruples about extending the content of a theory beyond what can be empirically supported. There is, on this view, empirical support for the basic equations of a theory but not for any causal claims that are not implied by the equations. As I will argue in chapter 7, however, such empiricist worries can be answered. Manipulability or interventionist accounts of causation show how our experimental interactions with otherwise closed systems—that is, with systems that can be treated as closed except for our experimental interactions with them—can provide empirical support for asymmetric causal assumptions not implied by a theory's fundamental dynamical equations. Our experimental interactions with electromagnetic systems suggest that changes in the trajectories of charged particles have an effect on the state of the field in the future but not in the past.

3. Theories and Their Domains

I will argue that classical electrodynamics of charged particles provides us with an example of inconsistent representations that appear to be 'here to stay,' since (as I will discuss in quite some detail) there appears to be no overall more successful, fully consistent alternative scheme for modeling classical phenomena involving charged particles. But, one might object, how can I claim that classical electrodynamics is here to stay? Has the theory not already been superseded by quantum electrodynamics (and are its foundational problems not perhaps of at best historical interest)? I want to reject the picture of theory change that is presupposed by this last question. Attempts at constructing a comprehensive classical electromagnetic *world picture* have been given up and have been superseded by a quantum mechanical view of the micro world. That is, classical electrodynamics has been abandoned as *global* representation of what the world fundamentally is like. But this does not mean that the theory has been abandoned *tout court*. Rather, classical electrodynamics is still regarded as providing us with the most successful models in the domain of classical phenomena involving charged particles, and the theory has continued to be one of the core theories of twentieth- (and twenty-first)-century physics.

Philosophers of science have frequently characterized theory change in science as involving the giving up of an old theory (after a sufficient number of the right sort of empirical difficulties for the theory have accumulated) and accepting a new, more promising rival. The major methodologists of science of the twentieth century, such as Karl Popper, Imre Lakatos, and Thomas Kuhn, thought of theory change in this way, and there seem to be many historical examples that fit this general characterization quite well. One stock example in the literature is the replacement of phlogiston theory by the oxygen theory of combustion. But the change from a classical electromagnetic worldview to a quantum conception of the world does not fit this model. In the case of phlogiston theory, the theory was in fact completely abandoned; yet in the case of classical electrodynamics the idea of a certain fundamental 'world picture' was given up, but not the belief that the theory was appropriate for modeling certain phenomena. Around the turn of the twentieth century, physicists proposed an electromagnetic worldview that was meant to replace the mechanical worldview of the eighteenth and nineteenth centuries. During the first decade of the twentieth century both empirical and foundational problems of the theory made it evident, however, that classical electrodynamics would not be able to fulfill the role of a fundamental and universal physics. Yet this realization did not lead physicists to abandon the theory as a means for constructing highly successful models of certain phenomena involving charged particles. Indeed, classical electrodynamics remains the most appropriate theory for phenomena in what we have come to understand as the 'classical domain.'

As Fritz Rohrlich puts it, classical electrodynamics is an "established theory" (see Rohrlich and Hardin 1983; Rohrlich 1988); that is, it is a theory that is empirically well supported within its domain of application, has known limits of empirical applicability, and coheres well with other theories. In particular, the theory coheres well with those theories, like quantum mechanics, which establish its boundaries. That classical electrodynamics coheres with quantum mechanics does not, of course, mean that the two theories are logically consistent with one another—they are not. Central to Rohrlich's notion of coherence is that the mathematical formalism of one theory can be derived from the other when the appropriate limits are taken. In the case of classical electrodynamics, the relevant limits include, as Rohrlich argues, a neutral-particle limit and quantum mechanical limits. The former is obtained by letting the magnitude of the charge go to zero. Rohrlich maintains that the theory should satisfy "the principle of the undetectability of small charges" (see Rohrlich 1990, 212–213), which requires that the equation of motion for a charged particle must approach that for a neutral particle as the charge goes to zero. The latter limit is characterized by the requirement that within the domain of classical physics, the value of classical and quantum mechanical observables must agree approximately, where the classical domain is characterized, roughly, by the condition that the value of classical observables is very large compared with the Planck constant h. Thus, on Rohrlich's view, which strikes me as correct, the development of quantum theory has not led to the demise of classical electrodynamics but, rather, has helped to determine the limits of the theory's domain of empirical applicability.

There is disagreement between those who think that the domains of application of *all* scientific theories are essentially limited and those who have a hierarchical conception of science according to which the ultimate aim of science is to discover some universal theory of everything. My claim that classical electrodynamics is (and to the best of our knowledge will remain) one of the core theories of physics is independent of that debate, and should be acceptable even to those who are hoping for a universal physics in the future. For, first, we need to distinguish between the *content* of a theory and our *attitude* toward that content. Even if we took the content of classical electrodynamics to be given by universal claims with unrestricted scope about classical electromagnetic worlds, our attitude can be that we accept or endorse those claims only insofar as they concern phenomena within the theory's domain of validity. Thus, the fact that the theory is false (and not even approximately true) if construed as having unrestricted scope does not force us to reject it outright. And, second, even if the actual world is fundamentally a quantum world (or one governed by a yet-to-be-discovered theory of quantum gravity), there are many phenomena that are best described in terms of classical electrodynamics. The latter theory is, and will remain, the most appropriate theory for modeling electromagnetic phenomena characterized by energy and length scales that are large compared with the Planck constant. Since a quantum treatment of many classical phenomena, if at all possible in practice, would be unduly complex without any significant gain in accuracy (given the energy scales involved), the classical theory scores highest for phenomena squarely in the classical domain, on some intuitive weighting of Kuhn's criteria, because its advantages in simplicity more than compensate for any extremely small losses in accuracy.

4. Overview of Things to Come

There are two basic issues concerning the interaction between charged particles and classical fields that I will discuss in this book: First, what is the equation of motion of a charged particle interacting with electromagnetic fields? And, second, how does the presence of charged particles, or sources, affect the total field? These two questions will, in some form or other, provide the overarching themes for the two halves of this book. The focus of chapters 2 through 4 will be on particles—on different particle equations of motion and their properties. The main focus of chapters 5 through 8, by contrast, will be on fields and their symmetry properties in the presence of particles.

In chapter 2, I show that the standard equation of motion for charged particles in an electromagnetic field is inconsistent with the field equations and energy conservation. Yet this inconsistency is not fatal to the success of the equation in modeling the motion of charged particles. I argue that 'logical anarchy' is avoided in the resulting theory since, first, implicit in the theory are certain guidelines or rules that constrain the application of its basic equations; and, second, different inconsistent predictions from the theory agree approximately within the theory's domain of application. The rules governing the application of the equations are naturally seen as arising from the theory's causal structure, in that the different models represent what are naturally thought of as different causal routes by which fields and particles interact.

Readers who have some familiarity with classical electrodynamics may find my claim that the theory is inconsistent surprising. Textbooks on classical electrodynamics do not treat the theory differently, in any obvious way, from other, arguably consistent theories. In fact, even most advanced textbooks do not make it easy for the reader to detect that the theory of charged particles I will discuss is indeed inconsistent. Now, it strikes me that the very fact that the theory is not treated as an anomaly, and that one can obviously learn much about the theory without ever realizing that it is inconsistent, is in itself philosophically illuminating. If inconsistencies posed a serious threat to the empirical applicability and testability of a theory and required special attention, then one would expect textbook treatments to be far more careful than they in fact are in drawing attention to the inconsistency of the point particle theory. Instead, a survey of standard advanced graduate- and research-level treatments (e.g., Jackson 1999 or Landau and Lifshitz 1951) suggests that one can learn how to apply the theory successfully, as well as what constraints govern the application of its fundamental equations, without being aware of the theory's foundational problems. Textbooks teach how to construct representational models of the phenomena in the theory's domain by teaching students to see these phenomena from certain perspectives, as it were. And it appears to be to a considerable extent irrelevant to the success of the theory whether the fundamental equations to which we appeal in constructing these models provide us with a consistent, unified account of possible ways the world could be. That is, it appears to be irrelevant to the success of the theory in enabling us to construct empirically successful models, whether the different perspectives presented by the theory cohere in such a way that they can be integrated into a single account of ways the world could be. I want to suggest that an adequate philosophical account of scientific theorizing ought to 'save this phenomenon.'

My discussion in chapter 2 of the role of inconsistent theories presupposes that the inconsistent set of equations on which I focus cannot be thought of as an approximation to some other, consistent classical theory. I argue for this assumption in chapter 3. There I examine several rival approaches to the standard, inconsistent formalism and discuss some of the difficulties one encounters in trying to arrive at a consistent classical theory of the interaction of particles and fields. Since a detailed discussion of several rather arcane electromagnetic theories may go beyond what nonspecialist readers are interested in, I decided to structure the presentation in this way. Readers who want to skip chapter 3 (or perhaps only skim section 3.3, on the Lorentz–Dirac equation, which has some relevance to the discussions in chapters 4 and 6) may want either to accept my claim that there is no conceptually unproblematic, consistent classical theory to which the standard particle theory is an approximation, or simply to read the conclusions of chapter 2 as conditional on that claim. As we will see, contrary to the particle theory, an electrodynamics of continuous charge distributions is in fact consistent. Hence one natural response to the inconsistency of the standard point particle scheme is to argue that the theory of classical electrodynamics fundamentally is a theory of extended distributions of charges and that the inconsistency is merely a consequence of introducing discrete particles into the theory. On this picture, then, the *theory* of classical electrodynamics is consistent, while what is inconsistent is only a certain *approximation* to it.

I argue, however, that the inconsistency is more central to theorizing in classical electrodynamics than this picture suggests. For one thing, a classical theory of continuous media *on its own* is inconsistent with the existence of discrete charged particles, and hence cannot govern phenomena involving charged particles even in principle. Thus, in order to be able to model the behavior of charged particles, one needs to postulate additional cohesive forces that can ensure the stability of finitely charged discrete objects. But there is no satisfactory and consistent dynamics for extended classical charged particles that takes into account these additional forces. Moreover, physicists do not seem to be all that concerned about this state of affairs. Existing inconsistent particle models suggest that a fully consistent theory, if it were to be possible at all, would have to be unduly complex. Thus, the history of classical charged-particle theories suggests that the aim for consistency can be overridden by the aim of arriving at a simple set of dynamical equations. Pragmatic considerations, such as that of mathematical tractability, enter even at the highest level of arriving at a theory's 'fundamental' equations.

Contrary to simple textbook cases of approximations, the standard theoretical scheme used to model the behavior of classical charged particles is not backed up by a complete theory which we could take to govern the relevant phenomena in principle. Or, put differently, in modeling phenomena involving microscopic charged particles, classical electrodynamics is applied to a domain for which the theory has no model-theoretic models. Thus, the particle theory is not so much an *approximation* to the continuum theory as an inconsistent *extension* of it. That a theory can be extended into a domain in which it has no model, I take it, is perplexing on standard views of theories, independently of whether we wish to take the theoretical scheme constituting the extension to be a theory in its own right, or whether we want to insist that only the consistent continuum theory deserves the honorific title 'theory.' In what follows, I shall for ease of exposition often call the Maxwell–Lorentz point particle scheme a 'theory,' but I hope that none of my arguments depend on that terminological choice.

I want to emphasize from the outset that my conclusion will not be that inconsistencies are completely unproblematic. Clearly, all else being equal, a consistent theory is better than its inconsistent rivals. And consistency may even be a more important desideratum for theories than some of the other demands we wish to place on our theories. But I suspect that cases where all else is equal are rare. And where the aim for consistency conflicts with some of the other demands we place on our theories, it may be the condition of internal consistency that is given up.

Chapter 4 focuses on the notions of locality and causation in classical particle–field theories. It is often said that one advantage of classical field theories is that they are 'local' and do not involve action-at-a-distance. While locality conditions, such as the condition that no force can act where it is not, often are informally introduced in causal terms, it nevertheless is a common view that there is no place for 'weighty' causal notions in fundamental physics. This view famously was advanced by Russell in the last century (Russell 1918), and has recently received renewed attention, partly due to a paper by Hartry Field (Field 2003), where Field approvingly presents Russell's claims, but also cites arguments due to Cartwright for causal realism, according to which 'weighty' causal assumptions are essential to effective strategizing (Cartwright 1979). According to Field, this tension presents an important metaphysical problem.

If Russell (and Field) were right, then any putatively causal locality condition would have to be either abandoned or given a noncausal analysis. In chapter 4 I distinguish several different locality conditions that have not been distinguished carefully in the literature and argue that two of these conditions are irreducibly causal. As my main case study I discuss the Lorentz–Dirac theory, which, despite the fact that it is a classical particle–field theory, is causally nonlocal. I argue that causal assumptions play an important role in the theory and that these assumptions can neither be reduced to claims about functional dependencies, as Russell thought, nor be easily abandoned.

The main topic of chapters 5 through 8 is the asymmetry of radiation. Phenomena involving electromagnetic radiation exhibit a temporal asymmetry: In nature there appears to be coherent radiation diverging from radiating sources but no (or almost no) coherent radiation converging on sources. This temporal asymmetry is seen to present a puzzle or problem, since the Maxwell equations which are used to model radiation phenomena are time-symmetric (in a certain sense). In chapter 5 I introduce some of the necessary formal background for discussing the asymmetry and discuss the question of what, exactly, the asymmetry of radiation is. Often the problem of the asymmetry is introduced through a discussion of the views of two historical predecessors—Albert Einstein and Karl Popper. I will follow this custom and will try to correct certain misunderstandings of Einstein's and Popper's views on the asymmetry of radiation. Both Einstein's and Popper's discussions of the subject, I argue, provide useful clues to a successful solution to the puzzle—an appeal to asymmetric elementary radiation processes coupled with statistical considerations. I also discuss and criticize two more recent purported solutions to the puzzle proposed by Fritz Rohrlich and James Anderson.

The most influential attempts at solving the puzzle of the arrow of radiation without invoking explicitly causal assumptions appeal to the role that absorbing media play in electromagnetic processes. Absorber theories of radiation try to relate the asymmetry either to a temporal asymmetry of thermodynamics or to one of cosmology. In chapter 6, I critically discuss several such attempts, which can be traced back to the absorber theory of radiation of John Wheeler and Richard Feynman, and argue that all these attempts fail.

In chapter 7, I first present what I take to be the most successful absorber account of radiation—an account due to H. Dieter Zeh. Unlike the accounts criticized in chapter 6, Zeh's account may actually succeed in deriving the asymmetry to be explained in a non-question-begging way. Nevertheless, I argue, there are good reasons for rejecting Zeh's account. I then present what I take to be the most promising solution to the puzzle of the arrow of radiation. This solution appeals to the *retardation condition*, according to which the field associated with a charged particle is a diverging wave. I argue that this condition is best thought of as a time-asymmetric causal constraint.

David Lewis has cited the asymmetry of wave phenomena as an example of an asymmetry of overdetermination, according to which the future massively overdetermines the past, but not vice versa. Lewis's thesis of the asymmetry of overdetermination is at the core of his counterfactual account of the asymmetry of causation. In chapter 8, I argue that Lewis mischaracterizes the asymmetry of radiation. The latter is not an asymmetry of overdetermination, and Lewis's thesis of overdetermination is false. Thus, Lewis's counterfactual analysis of causation is left without an account of the asymmetry of the causal relation. Lewis's thesis crucially relies on the claim that, roughly, possible worlds diverging from ours are closer to the actual world than possible worlds perfectly converging to it. I suggest a way in which Lewis might amend his similarity metric between possible worlds to save this claim. However, if I am right in maintaining that the asymmetry of wave phenomena is at heart a causal asymmetry, then even the amended similarity metric will not save Lewis's analysis of causation, since the account would become circular.

PART ONE

PARTICLES

2

Inconsistency

1. Introduction

In this chapter and the next I want to examine in some detail how interactions between charged particles and fields are modeled in classical electrodynamics. One of the upshots of my investigation will be that there is no fully satisfactory mathematical representation of classical systems of particles and fields. It appears that despite the indubitably great successes of classical particle–field theory, there are limits to the adequacy of its mathematical models that go beyond questions of empirical adequacy. Perhaps the most startling aspect of such models is that the main approach to treating the interactions between classical charged particles and fields relies on a set of internally inconsistent assumptions.

Among the criteria for evaluating scientific theories proposed by philosophers, internal consistency appears to be privileged. A highly successful theory may be more or less accurate or may be more or less simple, but according to what appears to be a widely held view, internal consistency is not similarly up for negotiation: Internal consistency is a necessary condition for a theory to be even minimally successful, for an inconsistent set of principles threatens to be trivial, in that any sentence whatever can be derived from it. Modeling in classical electrodynamics provides a counterexample to this view. The standard approach to modeling particle–field interactions is inconsistent yet empirically is strikingly successful. What is more, there does not appear to be an otherwise at least equally successful, yet fully consistent, alternative theory that this approach could be taken to approximate. Thus a theory, it seems, can be successful without presenting a coherent account of what is physically possible or delineating a coherent class of physically possible worlds. While internal consistency may be a particularly important criterion of theory evaluation, it is not a necessary condition, for the aim for consistency can conflict with other demands we might want to place on a successful theory; and where such conflicts occur, there may be good reasons for giving up on full consistency.

One of the upshots, then, of my discussion in this chapter and the next will be that the logicians' all-or-nothing condition of consistency is too blunt a tool to evaluate and compare competing approaches to classical electrodynamics. Another, related theme is that even at the level of what one might call *fundamental theory*, pragmatic considerations, such as the question of mathematical tractability, enter into the process of theorizing. This suggests, I claim, that it may not be possible to clearly delineate a foundational project of arriving at a coherent account of what the world may be like—a project for which consistency is often taken to be essential— from the more pragmatic enterprise of applying theories to represent particular phenomena. Finally, I will argue that if inconsistent theories can indeed play a legitimate role in science, then accounts of theory acceptance according to which accepting a theory implies a commitment to the truth of the theory's empirical consequences (for example, van Fraassen's construals of scientific realism and empiricism) have to be rejected. Instead of invoking the notion of truth, acceptance ought to construed as involving a commitment to a theory's reliability.

Section 2 introduces some of the main features of standard microscopic classical electrodynamics and serves as a general background to many of the issues that will occupy us throughout the book. In section 3, I derive the inconsistency of the theory. Section 4 is devoted to the question of how the theory can be successful despite its inconsistency. There are some recent accounts arguing that inconsistent theories can play a legitimate, yet limited, role in scientific theorizing as heuristic guides in theory development. I argue that these accounts—chief among them John Norton's—allow too limited a role for inconsistent theories. While inconsistent theories clearly have to satisfy certain constraints in order to be scientifically useful, the constraints proposed by Norton are far too restrictive. In their stead I propose a set of weaker constraints that can account for the success of classical electrodynamics.

Throughout my discussion I will refer to the scheme used to model classical particle–field phenomena as a 'theory.' Indeed, my discussion in section 4 depends on the assumption that there is no satisfactory consistent classical theory covering the same phenomena as the Maxwell–Lorentz theory. Indubitably some readers will object to this assumption and will argue that instead we ought to think of that scheme merely as an approximation to some other, consistent classical theory. I will address this worry in chapter 3, where I will discuss several alternative approaches to classical electrodynamics.[1] Most of these approaches, as we will see, are either inconsistent or otherwise conceptually deeply problematic. Yet there exists one consistent and, as it were, relatively well-behaved classical theory—the theory of continuous charge distributions. The problem with that theory, however, is that on its own it is inconsistent with the existence of discrete finitely charged objects and that there is no fully consistent way to complete the theory that would make it compatible with the existence of charged particles.

My conclusion—that there is no satisfactory complete and consistent theory which governs classical phenomena involving charged particles—is independent of whether we think of the most successful scheme for modeling such phenomena as a *theory* in its own right or merely as an *inconsistent extension* of the fundamentally consistent theory of continua to the domain of charged particles. On either view it is true that there is an important class of phenomena we can model with the help of a theory that has no model-theoretic models representing the phenomena. Despite our successes in constructing representational models of phenomena involving charged particles, there is no satisfactory theory that has models (in the logician's sense) involving charged particles.

2. Microscopic Classical Electrodynamics

The ontology of standard microscopic classical electrodynamics consists of two basic kinds of entities—microscopic charged particles, which are treated as point particles, and electromagnetic fields—and the theory describes how the states of particles and fields mutually determine one another. The basic laws of the theory that govern the interaction between charged particles and fields are the Maxwell–Lorentz equations, which in a standard three-vector notation can be written as follows:

$$\nabla \cdot \mathbf{E} = 4\pi\rho$$
$$\nabla \times \mathbf{B} - \frac{\partial \mathbf{E}}{\partial t} = \mathbf{J}$$
$$\nabla \times \mathbf{E} - \frac{\partial \mathbf{B}}{\partial t} = 0 \qquad (2.1)$$
$$\nabla \cdot \mathbf{B} = 0$$
$$\mathbf{F}_{Lorentz} = q(\mathbf{E}_{ext} + \mathbf{v} \times \mathbf{B}_{ext}).$$

Here boldface symbols represent vector quantities. $\mathbf{E}$ and $\mathbf{B}$ are the electric and magnetic field strengths, respectively; ρ and $\mathbf{J}$ are the charge and current density; $\mathbf{F}_{Lorentz}$ is the Lorentz force. For point charges, the charge density for point particles is represented mathematically by a δ-function.[2]

The theory is a *microscopic* theory. It is a direct descendant of theories developed by Hendrik A. Lorentz and others, who around the turn of the twentieth century tried to provide a microscopic basis for nineteenth-century Maxwellian electrodynamics, which is a macroscopic theory. Standard textbooks generally discuss both the microscopic theory and a modern version of the macroscopic theory, which aside from the macroscopic analogues of the microscopic electric and magnetic field strengths, involves two additional vector fields, $\mathbf{D}$ and $\mathbf{H}$, that depend on properties of the medium in which the fields propagate. Following Lorentz, one can show that a microscopic distribution of discrete charged particles constituting a material object, together with the microscopic fields associated with the charge distribution, give rise to macroscopic fields, where the latter are derived by smoothing over the discontinuous distributions of discrete charges of the micro theory through suitable averaging procedures (see, e.g., Jackson 1975, sec. 6.7). Not distinguishing carefully enough between a theory of *continuous* charge distributions and a theory of *discrete* compact localizations of charges is one of the reasons why the standard theory might on first glance look foundationally unproblematic (see chapter 3). The focus of this chapter and of most of this book is on theories of charged particles.

According to the Maxwell–Lorentz equations, charges and electromagnetic fields interact in two ways. First, charged particles act as sources of fields, as determined by the four microscopic Maxwell equations; and, second, external fields influence the motion of a charge in accordance with the Lorentz force law. According to the first Maxwell equation, Coulomb's law, electric charges act as sources of electric fields, while the fourth equation states that there are no magnetic charges. The second equation, Ampere's law (including the term for the *displacement current* $\partial E/\partial t$, which was added by Maxwell), says that a current and a changing electric field induce a magnetic field. The third equation, Faraday's law, states that a changing magnetic field induces an electric field. The two inhomogeneous Maxwell equations are equivalent to the principle of the conservation of charge, which is one of the fundamental assumptions of classical electrodynamics. On the one hand, the two equations imply the continuity equation

$$\frac{\partial \rho}{\partial t} = -\nabla \cdot \mathbf{J}, \qquad (2.2)$$

which states that charge is conserved locally. On the other hand, it follows from the continuity equation that one can define fields whose dependence on charge and current is given by the Maxwell equation.

In the standard interpretation of the formalism, the field strengths **B** and **E** are interpreted realistically: The interaction between charged particles is mediated by an electromagnetic field, which is ontologically on a par with charged particles and the state of which is given by the values of the field strengths. For many purposes it is convenient to introduce *electromagnetic potentials* in addition to the field strengths. Since the divergence of a curl vanishes, the fourth Maxwell equation is automatically satisfied if one defines **B** in terms of a vector potential:

$$\mathbf{B} = \nabla \times \mathbf{A}. \qquad (2.3)$$

Faraday's law is then satisfied identically if the electric field **E** is defined in terms of **A** and a scalar potential Φ:

$$\mathbf{E} = -\nabla \Phi - \frac{1}{c}\frac{\partial \mathbf{A}}{\partial t}. \qquad (2.4)$$

Plugging the electromagnetic potentials into the Maxwell equations allows one to derive the wave equation for the potentials with well-known standard solution techniques.

I have presented the Maxwell–Lorentz equations in a traditional three-vector notation. Since the Maxwell equations are invariant under Lorentz transformations, they can also be written in a four-vector notation which exhibits the relativistic invariance explicitly. If we define the four-current $J^\alpha = (c\rho, \mathbf{J})$, the four-potential $A^\alpha = (\Phi, \mathbf{A})$, and the electromagnetic field tensor $F^{\mu\nu} = \partial^\mu A^\nu - \partial^\nu A^\mu$, whose six independent components represent the electric and magnetic field strengths (and can be written as a 4×4 matrix), then the Maxwell equations become

$$\partial_\mu F^{\mu\nu} = \frac{4\pi}{c} J^\nu \qquad (2.5)$$
$$\partial^\lambda F^{\mu\nu} + \partial^\mu F^{\nu\lambda} + \partial^\nu F^{\lambda\mu} = 0,$$

where, according to the standard convention, all Greek indices range over the four space–time coordinates 0, 1, 2, 3. Both equations are invariant under Lorentz transformations. Thus, classical electrodynamics is most naturally understood as a relativistic theory. It is a *classical* theory only in that it is not a *quantum* theory. However, in what follows, I will often appeal to three-vector versions of various equations, since the notation will probably be more familiar to many readers and most of my discussion does not depend on exhibiting the relativistic invariance explicitly.

Finally, the expression for the energy stored in the electromagnetic field will be important in what follows. Electromagnetic fields are taken to carry both energy and momentum. The energy density of the electromagnetic field is defined as

$$u = (8\pi)^{-1}(\mathbf{E} \cdot \mathbf{E} + \mathbf{B} \cdot \mathbf{B}), \tag{2.6}$$

and the energy flow is given by the Poynting vector,

$$\mathbf{S} = c(4\pi)^{-1}(\mathbf{E} \times \mathbf{B}). \tag{2.7}$$

The corresponding relativistic quantity is the *energy–momentum tensor* $T^{\mu\nu}$, where $T^{00} = u$ and $T^{0i} = c^{-1}S_i$. Here the index i ranges over the spatial coordinates 1, 2, 3. I will discuss possible motivations for adopting these expressions for the field energy in chapter 3.

Intuitively, the Poynting vector represents the local flow of energy stored in the field. Yet this interpretation is not entirely unproblematic. For there are electrostatic situations in which the Poynting vector is different from zero at most points, even though intuitively it would seem that the energy flow should be zero in such cases, since the field remains constant through time. A simple example is that of a point charge constrained to sit next to a permanent magnet. The fields in this situation—the Coulomb field of the charge and the magnetic field of the magnet—do not change with time, and one would expect that there should be no energy flow, yet (for almost all possible configurations) **E** and **B** are not parallel at most points in space and **S** does not vanish. However, for any *closed* surface the integral of the Poynting vector over the surface is zero in this case. That is, the total energy associated with any arbitrary volume remains constant. This suggests an interpretation according to which we should not think of energy as some kind of 'stuff.' If we interpret energy as a *property* of the electromagnetic field—or more specifically of the field in any arbitrary region of space—then the Poynting vector can be understood as specifying changes in this property. On this interpretation only the net flow across a closed surface, but not the local flow in or out of the enclosed volume, receives a physical interpretation. See Lange (2002, 152) for a defense of this interpretation.

The electromagnetic field associated with a charge can be broken into two components. The first component is a generalized Coulomb field, which does not carry energy away from a charge, since the total energy flow is given by integrating the Poynting vector over some surface enclosing the charge and this integral is zero if the field is a pure Coulomb field. Thus, the Coulomb field in a sense remains 'attached' to the charges with which it is associated. The second component is a radiation field associated with accelerated charges. The radiation field propagates at a finite speed, carrying energy away from the charge, and is understood to exist independently of the charge.

The temporal direction of electromagnetic radiation does not follow from the Maxwell equations alone, which are time-symmetric, but in textbooks is usually introduced as a separate assumption via the *retardation condition*. The Maxwell equations for the fields or potentials associated with a point charge have two solutions: The first, the *retarded solution*, represents a field diverging from the source; the second, the *advanced solution* represents a field converging into the source. The latter solution is usually rejected as unphysical or 'noncausal,' while the former can be given an intuitively straightforward causal interpretation: The acceleration of a charge is understood to be the cause of a local excitation of the electromagnetic field, which (in the usual time direction) propagates outward and away from the charge with finite speed. In chapter 7, I will defend the legitimacy of the notion of the field associated with a charge and will argue for a causal interpretation of the retardation condition.

The effect of external electromagnetic fields on charged particles is given by the Lorentz force law, which is treated as a Newtonian force law. That is, the expression for the force is used as input in Newton's second law, according to which the momentum change of the charge is equal to the total external force acting on it. If the only force present is the Lorentz force, then $dp/dt = \mathbf{F}_{Lorentz}$ or, in a relativistic formulation of the theory, $dp^\mu/d\tau = F^\mu_{Lorentz}$. Thus, even though Maxwell–Lorentz electrodynamics is most properly formulated in a relativistically invariant way, the theory is a Newtonian theory in one important respect: In order to arrive at an equation of motion for a charged particle, one needs to invoke the relativistic analogues of Newton's laws of motion. (Hence, when I speak of Newton's laws, I intend this to include their relativistic generalization and do not mean to draw a contrast between nonrelativistic and relativistic physics.) For the sake of simplicity, by the *Lorentz force law* I will mean not only the expression for the force but also the entire Newtonian equation of motion.

The interaction governed by the Lorentz force is local, in that the acceleration of a charge depends on the value of the electromagnetic field only at the location of the charge. As in the case of mechanical forces, the association between external fields and the acceleration of a charge is usually interpreted causally: External fields cause charges to accelerate.

Different charges, then, interact via the electromagnetic fields with which they are associated. Each charge gives rise to a field, which in turn affects the motion of all other charges.

I said that the electromagnetic field strengths are interpreted realistically on the standard interpretation of the theory. That is, electromagnetic fields are taken to be ontologically on a par with charged particles and, together with the latter, constitute the inventory of the world, as it were. There appear to be three types of argument for interpreting fields realistically. First, a theory with fields satisfies various locality conditions, such as the condition that causal interactions between charged particles are not transmitted across 'gaps' in space and time. An electrodynamics without fields would be guilty of the great "absurdity" (Newton, quoted in Lange 2002, 94) of allowing one body to act upon another at a distance, troubling Newtonian gravitational theory.

The second argument for interpreting fields realistically appeals to our commitment to the principles of energy and momentum conservation. Since electromagnetic effects are transmitted at a finite speed, energy and momentum would not be conserved in the interactions between charges in a version of the standard theory without fields. In the standard theory we assume that the interaction between charged particles is fully retarded—that is, we assume that the force one charge exerts on another is transmitted at a finite speed and that the force acting on a charge depends on the positions and the states of motion of all other charges at *earlier, retarded* times. This has the consequence that the force two charges exert on one another will not in general satisfy Newton's third law of being equal in size and opposite in direction. Thus, momentum is not conserved in interactions between charged particles and, hence, neither is energy.[3] (Notice that this difficulty does not arise in an *instantaneous* action-at-a-distance theory like Newtonian gravitational theory.) The problem can be avoided in electromagnetism if we introduce the electromagnetic field as carrier of real energy and momentum. The energy and momentum change of a charge can then, in principle, be balanced locally by a change in field energy and momentum. As we will see in the next section, however, even with real fields the Lorentz force law and the Maxwell equations are inconsistent with energy conservation, and energy is only approximately conserved in the standard theory.

A third consideration that may favor real fields is that there is no well-posed initial-value problem for a relativistic pure-particle theory, since specifying the positions and velocities of all charged particles on a spacelike hypersurface does not determine the dynamical motions of the charges. If influences from one charge on another propagate at a finite speed, the motion of a given charge in the future of the initial-value surface is partially determined by the motions of other charges in the past of that surface, where the latter motions are not determined by data on the initial-value surface. By contrast, since interactions between particles are mediated by fields in a field theory, the influences of past particle motions on a given charge are encoded in the state of the field on any given hypersurface. Thus, one might suspect that a field theory allows, at least in principle, the formulation of an initial-value problem. As we will see, however, only for a theory of fields and *continuous* charge distributions is there a well-defined initial-value problem, and even in this case there is a wide class of physically reasonable initial conditions for which there are no global solutions.

All three motivations for interpreting the fields realistically can be understood as attempts to render the theory 'well-behaved' in one sense or other. We will see, however, in this chapter and in chapters 3 and 4 that, ironically, these attempts ultimately are not fully successful. Even with real fields, there is no version of the theory that fully conforms to the classical methodologist's wish list.

Given the value of any free incoming fields, the electromagnetic field in a certain region is completely determined by the charges and currents in that region. The electromagnetic potentials, on the other hand, are not uniquely defined by the field strengths, but only up to what is known as a *gauge transformation*: The magnetic field **B** and the electric field **E** are left unchanged by the simultaneous transformations $\mathbf{A} \to \mathbf{A}' = \mathbf{A} + \nabla \Lambda$ and $\Phi \to \Phi' = \Phi - (1/c)(\partial \Lambda/\partial t)$, for some scalar function Λ. Thus, since the potentials are radically underdetermined by what is empirically measurable, they are traditionally not taken to have any real physical significance and are understood to be pragmatically useful excess mathematical structure.

This traditional interpretation of real fields and unreal potentials has been challenged in various ways. On the one hand, Wheeler and Feynman (1945) have argued that energy conservation can be compatible with a pure particle version of electrodynamics without fields, if the interaction between charged particles is symmetric and forces between charged particles propagate along both future and past light cones. On the other hand, Belot (1998) and Healey (2001) cite arguments appealing to the quantum-mechanical Aharanov–Bohm effect (which may suggest a realist interpretation of electromagnetic potentials), and considerations of interpretive continuity between classical and quantum theories, to support a realist construal of the potentials even in a classical context. I will discuss Belot's views and the status of various locality conditions in classical electrodynamics in detail in chapter 4. Wheeler and Feynman's absorber theory will be the focus of chapter 6.

The mutual interactions between charges and fields in Maxwell–Lorentz electrodynamics satisfy several demands one might intuitively wish to place on a causally well-behaved theory. For one, the interactions are causally local in two distinct senses: The influence of one charge on another is transmitted at a finite speed; and, due to the presence of the electromagnetic field, effects are not transmitted across spatiotemporal 'gaps.' In addition, the theory including the retardation condition satisfies the condition that effects do not precede their causes. Thus, classical electrodynamics fits extremely well into a causal conception of the world, and in fact is often taken to be the paradigm of a local and causal physical theory. Moreover, the theory is predictively extremely accurate within the domain of classical physics and the Maxwell–Lorentz equations are often said to satisfy certain intuitive criteria of simplicity. The theory, that is, scores very high on a number of criteria of theory assessment, such as conceptual fit, accuracy, and simplicity. Yet, as I want to show next, it fails miserably on what may appear to be the most important demand—the theory is mathematically inconsistent.

3. The Inconsistency Proof

The Maxwell–Lorentz equations allow us to treat two types of problems (see Jackson 1975, 1999). We can use the Maxwell equations to determine the fields associated with a given charge and current distribution, or we can use the Lorentz force law to calculate the motion of a charged particle in a given external electromagnetic field. In problems of the first type, the charges and currents are specified and, given particular initial and boundary conditions (which specify the source-free fields), the total electromagnetic field is calculated. In problems of the second type, the external electromagnetic fields are specified and the motions of charged particles or currents are calculated. Electric charges are treated *either* as being affected by fields *or* as sources of fields, but not both. That is, in both types of problems one ignores any effects that the field associated with a charge itself—the *self-field*—might have on the motion of that charge.

One can also, in a stepwise treatment, combine the two types of problems. An example is models of synchrotron radiation, which is the radiation emitted by circularly accelerated electrons in the magnetic field of a synchrotron accelerator (see Jackson 1975, sec. 14.6). In a first step, the orbit of the electrons in the external magnetic field is calculated using the Lorentz force law, ignoring the electrons' self-fields and, hence, any radiation. In the simplest model of a synchrotron, the electrons are assumed to be injected at right angles into a constant, purely magnetic field. In that case the Lorentz force equation of motion implies that the electrons move in a circular orbit. In a second step, the trajectories of the electrons are assumed to be given and are used as input to calculate the radiation field.

It is not difficult to see, however, that this stepwise treatment is inconsistent with the principle of energy conservation. On the one hand, since the electron orbit (as calculated in the first step) is circular, the electrons' speed and, hence, their kinetic energy are constant. (Moreover, since in this simple case the field is static, we can assign a potential energy to the electrons, which is constant as well.) On the other hand, charges moving in a circular orbit accelerate continuously (since the *direction* of their velocity changes constantly) and, thus, according to the Maxwell equations and the standard formulation for the field energy, radiate energy. But if energy is conserved, then the energy of the electrons has to decrease by the amount of the energy radiated and the electrons' orbit could not be the one derived from the Lorentz force in the first step, for which the electrons' energy is constant.

According to the Lorentz force law, the energy change of a charge is due only to the effects of external forces. Thus, since the external force in the synchrotron due to the magnetic field is at right angles to the charge's velocity, the speed of the electrons, and hence their kinetic energy, remain constant. Yet the charges radiate energy. Intuitively, this means that, for energy to be conserved, the electrons should accelerate less than similar neutral particles would. That is, contrary to what the Lorentz law predicts, we would expect the electrons to slow down.

Models of synchrotron radiation offer one illustration of how the theory's inconsistency manifests itself. I now want to derive the inconsistency more generally. The following four assumptions of the Maxwell–Lorentz theory are inconsistent:

(i) There are discrete, finitely charged accelerating particles.
(ii) Charged particles function as sources of electromagnetic fields in accord with the Maxwell equations.
(iii) Charged particles obey Newton's second law (and thus, in the absence of nonelectromagnetic forces, their motion is governed by the Lorentz force law);
(iv) Energy is conserved in particle–field interactions, where the energy of the electromagnetic field and the energy flow are defined in the standard way. (See equations (2.6) and (2.7).)

It follows from the Maxwell equations, in conjunction with the standard way of defining the energy associated with the electromagnetic field, that accelerated charges radiate energy, where the instantaneous power radiated is given by $P = 2e^2a^2/3c^3$ (Jackson 1975, 659). Thus, if the acceleration of a charge is nonzero at any time t, $t_A < t < t_B$, then the energy E_{rad} radiated by the charge between times t_B and t_A is greater than zero.

Newton's second law and the definition of the external work done on a charge imply that the work done on a charge is equal to the change in the energy of the charge. That is, (iii) implies that

$$W = \int_A^B \mathbf{F}_{ext} \cdot d\mathbf{l} = \int_{t_A}^{t_B} \frac{d\mathbf{p}}{dt} \cdot \mathbf{v} dt = m \int_{t_A}^{t_B} \frac{d\mathbf{v}}{dt} \cdot \mathbf{v} dt = \frac{m}{2}\left(v(t_B)^2 - v(t_A)^2\right) \quad (2.8)$$
$$= E_{kin}(t_B) - E_{kin}(t_A).$$

But for energy to be conserved, that is, for (iv) to hold, the energy of the charge at t_B should be less, by the amount of the energy radiated E_{rad}, than the sum of the energy at t_A and the work done on the charge. That is,

$$E_{kin}(t_B) = E_{kin}(t_A) + W - E_{rad}. \quad (2.9)$$

Equations (2.8) and (2.9) are inconsistent with one another if E_{rad} is finitely different from zero. But $E_{rad} > 0$ is implied by the conjunction of (i) and (ii). Thus, the core assumptions of the Maxwell–Lorentz approach to microscopic particle–field interactions are inconsistent with one another.

The argument for the inconsistency of the standard theoretical scheme for modeling electromagnetic phenomena involving charged particles is independent of the nature of the external force acting on the charge, yet it will be useful to consider the special case of a charged particle interacting with an external (arbitrary) electromagnetic field and no nonelectromagnetic forces, since variants of the following derivation will play an important role later. The principle of energy–momentum conservation for the entire system consisting of the electromagnetic field and the charge requires that, for a certain volume V containing the charge, any change in the particle's energy should equal the change in the field energy within V plus any field energy flowing across the boundary of V. Or, equivalently, the net rate at which field energy is flowing across the boundary of V should equal the rate at which the field energy in V is decreasing, minus the rate at which the field is doing work on the charged particle:

$$\oint \mathbf{S} \cdot d\mathbf{a} = -\frac{d}{dt}\int_V u d^3x - \frac{d}{dt}E_{charge}, \quad (2.10)$$

where $\mathbf{S}$ and u are the Poynting vector and the energy density, respectively, introduced above. However, this conservation principle is not satisfied in the electrodynamics of discrete charged particles governed by the Lorentz force law.

To see this, we can begin with the left-hand side. The surface integral giving the rate of energy flow can be transformed into a volume integral using Gauss's law:

$$\oint \frac{1}{4\pi c}(\mathbf{E} \times \mathbf{B}) \cdot d\mathbf{a} = \int_V \frac{1}{4\pi c}\nabla \cdot (\mathbf{E} \times \mathbf{B}) d^3x. \quad (2.11)$$

This expression can be further transformed with the help of the Maxwell equations, which imply that the right-hand side of (2.11) is equivalent to

$$-\frac{\partial}{\partial t}\int_V \frac{\mathbf{E}\cdot\mathbf{E}+\mathbf{B}\cdot\mathbf{B}}{8\pi}d^3x - \int_V \mathbf{J}\cdot\mathbf{E}d^3x. \qquad (2.12)$$

The first term can be identified with the change of the energy stored in the field in a volume V, according to the definition for the field energy introduced above. The problematic term is the second term: The field coupled to the current **J** is the *total* field, including both the external field and the field due to the charge itself, yet, according to the Lorentz equation of motion, the rate of work done on the charge is $q\mathbf{v}\cdot\mathbf{E}_{ext}$ and involves only the *external* field. If we express the current density of the charged particle with position $\mathbf{r}(t)$ with the help of a δ-function as $\mathbf{J}(\mathbf{x},t) = e\mathbf{v}(t)\delta[\mathbf{x}-\mathbf{r}(t)]$, then the energy change of the charge implied by the Lorentz force law is given by

$$\frac{dE_{charge}}{dt} = \int_V \mathbf{J}\cdot\mathbf{E}_{ext}d^3x. \qquad (2.13)$$

Finally, we can plug (2.13) into (2.12), after we split the total electric field in (2.12) into two components, a self-field due to the charge and an external field. The result is

$$\oint \mathbf{S}\cdot d\mathbf{a} = -\frac{\partial}{\partial t}\int_V u d^3x - \int_V \mathbf{J}\cdot\mathbf{E}_{ext}d^3x - \int_V \mathbf{J}\cdot\mathbf{E}_{self}d^3x$$
$$= -\frac{d}{dt}(E_{field}+E_{charge}) - \int_V \mathbf{J}\cdot\mathbf{E}_{self}d^3x. \qquad (2.14)$$

Therefore, the energy flow out of the volume is not balanced by the change in the field energy and the work done on the charge, as the principle of energy conservation (2.10) would require. Instead there is an extra term, involving the charge's interaction with its own field.

As this discussion suggests, the inconsistency is most plausibly seen as arising from the fact that the Lorentz force equation of motion ignores any effect that the self-field of a charge has on its motion. The standard scheme treats charged particles as sources of fields and as being affected by fields—yet not by the total field, which includes a contribution from the charge itself, but only by the field external to the charge. This treatment is inconsistent with energy conservation. Intuitively, if the charge radiates energy, then this should have an effect on its motion, and thus a radiation term representing a force due to the charge's self-field should be part of the equation of motion. That is, (2.13), which is implied by the Lorentz force equation of motion, should be amended. But, as we will see in chapter 3, it appears to be impossible to add a radiation term to the equation in a way that is both consistent and conceptually unproblematic.

4. Inconsistency and Theory Acceptance

4.1. Theories as Mathematics Formalisms

The inconsistency of the standard approach to modeling classical electromagnetic phenomena raises the following obvious puzzle: If drawing inferences from the theory's fundamental equations is anything like standard deductive inference, then we should be able to derive any arbitrary sentence from the inconsistent fundamental equations. The theory would be trivial. So how is it that the theory nevertheless appears to have genuine empirical content? To be sure, derivations in mathematical physics are not always easily reconstructed as logical deductions in a formal language. But clearly they proceed 'quasi-formally' within some appropriate mathematical language, and consistency appears to be an important constraint on such derivations. In particular, if a theory allowed us to derive inconsistent predictions, it is not clear what could count as either confirming or falsifying evidence for the theory. Suppose the theory allowed us to make two inconsistent predictions O_1 and O_2, such that the truth of O_1 implied the falsity of O_2, and O_1 turned out to be true. Would this count as confirming evidence for the theory, since O_1 is true, or as disconfirming evidence, since O_2 is false?

For some such reasons the view that inconsistency is a necessary condition for theories is widely held. To mention just a few examples, Pierre Duhem maintained, even as he was arguing for the view that there are only a few restrictions on the choice of hypotheses in physics, that "the different hypotheses which are to support physics shall not contradict one another. Physical theory, indeed, is not to be resolved into a mass of disparate and incompatible models; it aims to preserve with jealous care a logical unity" (Duhem 1962, 220). And Poincaré worried, concerning the use of both classical and quantum ideas in derivations in the old quantum theory, that "there is no proposition that cannot be easily demonstrated if one includes in the demonstration two contradictory premises" (quoted in Smith 1988, 429 fn.). Similarly, Karl Popper (1940) held that accepting inconsistent proposals "would mean the complete breakdown of science." As we have seen in the introduction, the prohibition of inconsistency is also built into standard accounts of the structure of scientific theories, such as the syntactic view, which identifies theories with deductively closed sets of sentences, and those versions of the semantic view which identify theories with a coherent class of models delineating the possible worlds allowed by the theory.

Classical electrodynamics appears to present a problem for the reigning orthodoxy. Or does it? One may want to object to my inconsistency proof by pointing out that I have not shown that the formalism of the theory on its own is inconsistent. Rather, what I have shown is that the conjunction of the claims that there are finitely charged particles and that the behavior of these particles is governed by the equations in question is inconsistent. But this does not, of course, imply that the formalism by itself has no (model-theoretic) models. This can be seen by looking at equation (2.14). If there is a domain of objects for which the last term on the right-hand side, the self-energy term, is zero (or makes no finite contribution), then the theory has a model. And in fact, as we will see in the next chapter, there is such a domain—that of continuous distributions of infinitesimally charged particles or *charged dusts*, for which the self-fields contribute only infinitesimally to the energy balance. Thus, if we were to identify the theory of classical electrodynamics with its mathematical formalism *without including claims about ontology*, then the theory would be consistent.

The assumption behind this objection is the idea that existence claims are not part of a scientific theory proper. But this assumption strikes me as misguided.

Consider, for example, an imaginary theory that postulates *both* that all massive bodies attract each other according to Newton's $1/r^2$-law of universal gravitational attraction *and* that all massive bodies exert a $1/r^3$-force on each other. On their own—that is, without the further assumption that there are at least two massive bodies—the two seemingly contradictory force laws and Newton's laws of motion form a consistent set, for the theory has models: universes containing at most one massive object. Yet if this theory is intended to be a theory of multiparticle interactions, then it appears to me that the correct way to describe it is as inconsistent. While the claim that there are at least two massive objects arguably is not fundamental or lawlike, it appears to be an integral part of any theory of multiparticle interactions. Scientific theories are *about* certain things, and a claim stating that what a given theory is about exists, ought to be considered part of that theory.[4] As *a theory of multiparticle interactions*, a theory with incompatible force laws is inconsistent.

Theories provide us with representations of the phenomena in their domain, and a theory's representational content may go beyond what is captured in the theory's fundamental dynamical equations. For example, a theory may account for a phenomenon by positing that the world is populated by certain entities whose interactions give rise to the phenomenon in question. In fact, Maxwell–Lorentz electrodynamics, which is a direct descendant of Lorentz's attempts to derive Maxwell's macroscopic particle–field equations from the interactions of microscopic 'electrons' with microscopic fields, is just such a theory: Like Lorentz's theory, the modern theory treats many electromagnetic effects as ultimately due to the interaction of discrete, microscopic charged particles and fields. The claim that charge is 'quantized'—that is, that there are discrete particles that carry a charge equal to multiples of a basic unit of charge identical to the charge of an electron—is one of the central tenets of microscopic classical electrodynamics. In light of the central importance of the concept of discrete charged particles to post-Maxwellian classical electrodynamics, the fact that a view implies that the theory denies the existence of such particles (as the theory would, if it were to be identified with the set of its fundamental equations) strikes me as a reductio of that view. Just as molecular biology makes claims about the existence of DNA, microscopic electrodynamics postulates discrete charged particles. In both cases the existence claim is an integral part of the theory's content.

Moreover, even if we were to accept that existence claims cannot be part of the content of scientific theories, it does not solve the substantive problem at issue. The puzzle presented by classical electrodynamics is how it can be that a theory or a set of equations can be used to represent the phenomena in a certain domain, despite the fact that the theory's basic equations have no models in that domain. Those who would want to insist that consistency is a privileged criterion of theory choice would equally want to insist, I take it, that for a theory successfully to represent a certain range of phenomena, the theory would have to have models (in the logician's sense of structures in which the equations are jointly true) that can function as representations of these phenomena. Now, equations (2.1), together with the principle of energy conservation, imply that there are no charged particles. How is it, then, that we nevertheless can successfully represent electromagnetic phenomena in terms of charged particles that are governed by just this set of equations? At least one of the intended domains of the theory (and arguably its *main* domain) is that of discrete charged charged particles and fields. Yet the formalism has no models for that domain. That is, to the extent that the theory is meant to be a theory of charged particles, the class of the theory's *intended* model-theoretic models is empty—there are no structures within the theory's intended domain of which the theory's fundamental equations are true. Nevertheless, the theory is empirically successful within that domain. Therefore it must be possible to construct representations—or representational models—of the phenomena in that domain. Thus, the puzzle is how the theory can provide representations of which the theory's laws are not true. In other words, the question is how the theory can provide *representational* models which are not (in some loose sense) *model-theoretic* models of the theory. And this puzzle remains the same, independently of whether we want to say that the theory in question (taken to include the claim that there *are* charged particles) is inconsistent or whether we want to insist that the theory without the existence claim is consistent, yet has no models involving charged particles.

Should we perhaps think of discrete particles as an idealization within a consistent theory of continuous charge distributions? We tend to think of idealizations as computationally useful yet as 'leading us away from the truth.' Yet if the present suggestion is right, classical electrodynamics presents us with a case where introducing an idealizing assumption inconsistent with the fundamental equations of the theory dramatically improves the theory's predictive power and accuracy. As we will see in the next chapter, the laws of the continuum theory have no (model-theoretic) models that can *even in principle* adequately represent the behavior of compact localizations of charge, since they do not on their own allow for localized electromagnetic objects that retain their integrity through time. Thus, the theory quite dramatically fails empirically, while introducing the particle idealization leads to an empirically quite successful scheme. If, therefore, we assume that the continuum theory is the basic or fundamental theory, from which the particle approach is obtained by introducing the concept of discrete particles as idealization, then we are forced to conclude that there are idealizations—and even idealizations inconsistent with the theory's fundamental dynamics—that are absolutely essential to the theory's empirical adequacy.

4.2. A Role for Inconsistent Theories

One strategy for allowing for the possibility of inconsistent theories that are not trivial is to try to reconstruct scientific inferences in a *paraconsistent* logic. A paraconsistent logic is a logic that does not validate the inference schema *ex falso quod libet*—that is, inferences from sentences A and ~A to an arbitrary sentence B. Some advocates of paraconsistent logics hold the radical view that there are true contradictions in nature, which may even be observable (e.g., Priest 2002), while others propose alternatives logics as a way to model reasoning that involves inconsistent theories that are taken to be at best approximately true (e.g., Brown 2002; da Costa and French 2002). I doubt, however, that trying to distinguish derivations allowable in classical electrodynamics from those that are not, on purely formal grounds, is a promising strategy.

Here I would like to cite just two examples, which I take to be symptomatic of the problems faced by such formal accounts. One of the main approaches to paraconsistent logic proceeds by constructing a logic that is not adjunctive, which may be achieved by defining the consequence relation as follows: "A is a *consequence* of a set of formulas, Γ, exactly if A is classically implied by some subset in every partition of Γ into n nonoverlapping consistent subsets (where n is the minimum number of nonoverlapping consistent subsets into which Γ can be divided)." However, this condition appears to exclude some allowed inferences in classical electrodynamics. It seems plausible that the minimum number of nonoverlapping consistent subsets into which classical electrodynamics can be divided is two: Of the three fundamental principles of the theory I have identified—the Maxwell equations (M), the Lorentz force law (L), and energy conservation (E)—any two are mutually consistent, but cannot consistently be conjoined with the third (if we assume that there are charged particles—a premise I suppress here for ease of exposition). That is, the set of partitions is $\{\{M, L\}, \{E\}; \{M\}, \{E, L\}; \{M, E\}, \{L\}\}$. Yet models of phenomena in which charged particles are treated as both affected by external fields and as contributing to the total field—such as those modeling synchrotron radiation—can be derived only from both the Lorentz force law and the Maxwell equation, and thus are not implied by some subset in every partition. (They are not implied by any subset in the second and third partitions.) Moreover, while some derivations appealing to the Maxwell equations and energy conservation are licensed in the standard theory, others (namely those which make claims about the motion of a charge in an electromagnetic field) are not. Thus, we could not simply weaken the definition of the consequence relation such that A is a consequence of Γ, if it is classically implied by at least one consistent subset of Γ.[5] All the allowable inferences in classical electrodynamics proceed from some consistent subset of the theory's fundamental equations. But which of those are allowable and which are not seems to depend on the content of the basic laws, and cannot be determined formally alone.

As a second example, consider the partial structure approach outlined by Newton da Costa and Steven French (2002, 112). They argue that scientific theories ought to be represented as structures of the form $M = \langle A, R_i \rangle$ where A is a nonempty set and R_i is a family of relations. These structures are *partial* in that any relation of arity n need not be defined for all n_i-tuples of elements of A. Applied to our case, the partial structures approach would appear to recommend that different regions of space-time which contain systems of particles and fields (i.e., different subsets of A) satisfy some of the fundamental equations of the theory, but not all of them: Some particle–field systems satisfy the Lorentz equation of motion, while others satisfy the Maxwell equations and energy conservation, say. But this misconstrues the commitment scientists appear to have to the theory. It is not the case that we take some electrons to be governed by the Lorentz force equation and others by the Maxwell equations—our commitment to the approximate truth or, perhaps, 'van Fraassen-style' empirical adequacy of the Maxwell–Lorentz equations, extends to *all* classical systems of charges and fields. Nevertheless, for a given system we use only a proper subset of the theory's equations to model its behavior, where the choice of equations *depends on what aspect of the interaction between charges and fields we are interested in*. We can represent either the effects of electromagnetic fields on charged particles or the effects of particles on the fields, but, given energy conservation, not both simultaneously.

A general worry about formal approaches to dealing with inconsistencies of science is that even if there were some formal framework in which the allowable derivations in a given theory could be reconstructed, any argument for the adequacy of the framework would have to piggyback on an informal 'content-based' assessment of which inferences are licensed by the theory and which are not. Thus, even if the challenge of finding an adequate formal framework could be met, an additional challenge would be to show that the framework can add to our understanding of the theories in question. This challenge appears particularly daunting in light of the fact that scientists themselves do not seem to invoke any formal reasoning principles in a logic weaker than classical logic. If there is any formal constraint on theorizing to which scientists explicitly appeal on occasion, it is that of consistency!

There also have been challenges to the reigning orthodoxy from philosophers who do not advocate adopting an alternative logic.[6] John Norton has argued for what he calls a "content driven control of anarchy" (Norton 2002, 192). He maintains that Newtonian cosmology (Norton 2002)[7] and the old quantum theory of blackbody radiation (Norton 1987) are inconsistent. His diagnosis in both cases is that anarchy is avoided, since scientists employ constraints that selectively license certain inferences, but not others, on the basis of reflections on the specific content of the theory at issue. I believe that such a content-driven approach to reconstructing theorizing in the presence of inconsistencies points in the right direction, yet Norton's proposal regarding what kind of role inconsistent theories can play in science provides a relatively conservative revision of the traditional wholesale prohibition against inconsistencies, and here I disagree with him.

According to Norton, there are two fairly restrictive constraints on the permissibility of inconsistencies in theorizing. First, he holds that when physicists use an inconsistent theory, their commitment can always be construed as extending only to a *consistent subset* of the theory's consequences. That is, according to Norton, it always will be possible to reconstruct a permissible inconsistent theory by 'surgically excising' the inconsistency and replacing the theory with a single consistent subset of its consequences in all its applications. And second, for Norton an inconsistent theory is permissible only as a *preliminary stage* in theorizing that eventually is replaced by a fully consistent theory. Classical electrodynamics, however, does not fit either of these constraints. Thus, I want to propose a set of alternative constraints weaker than Norton's that nevertheless are strong enough to safeguard against logical anarchy and that can help us account for theorizing in electrodynamics.

As far as the first constraint is concerned, Norton holds that a theory's inconsistency is no threat to the theory's empirical applicability only if there is a consistent subset of the theory's consequences that alone is ultimately used to make empirical predictions. Take the old quantum theory of radiation, which appears to be prima facie inconsistent because it involves principles from classical electrodynamics and a quantum postulate. According to Norton, we can give a consistent reconstruction of the theory which consists of only a subset of classical electrodynamics together with the quantum postulate. In generating predictions, only this consistent subtheory is involved. In the case of Newtonian cosmology, we can derive from Newtonian gravitational theory, together with the assumption of Newtonian cosmology, that infinite Euclidean space is filled with a homogeneous isotropic matter distribution, that the net force on a test mass can have any magnitude and direction whatever (Norton 2002). Hence the postulates of the theory allow us to derive inconsistent conclusions. Yet, Norton maintains, scientists avoid being committed to inconsistent predictions by accepting only one of the infinitely many inconsistent force distributions derivable from the theory (2002, 191–192). This reconstruction of the attitude of scientists toward inconsistent theories echoes that of Joel Smith, who in a discussion of blackbody radiation and Bohr's theory of the atom says, "We find scientists choosing certain mutually consistent implications of otherwise globally inconsistent proposals to use together in the continuing inquiry" (Smith 1988, 444–445).

A similar reconstruction is not possible in the case of classical electrodynamics, since scientists endorse consequences of the theory which are mutually inconsistent, given the basic postulates of the theory, while also accepting these postulates in their entirety. As we have seen, scientists use the theory to make predictions based on the Lorentz force law, on the one hand, and predictions based on the Maxwell equations, on the other, without abandoning their commitment to the principle of energy conservation (which itself is invoked in certain derivations). Thus, there is no single consistent subset from which all the theory's acceptable empirical consequences can be derived. Rather, in different applications, scientists appeal to different internally consistent yet mutually inconsistent subsets of the theory's postulates. In contrast to the examples Norton and Smith discuss, a consistent reconstruction of the theory's entire predictive content is impossible. That means that one obvious route for 'sanitizing' inconsistent theories is blocked. This, however, raises the following problem: How can scientists be committed to incompatible predictions derivable from the theory, given that knowingly accepting inconsistent empirical consequences seems to be prohibited by standards of rationality?

One response to this problem would be to argue that we should revise our standards of rationality in a way that allows for knowingly accepting inconsistent claims. But I think nothing that radical is needed in the present case. Instead, I want to suggest that the source of the problem is a certain picture of *theory acceptance* that we should give up. The problem arises if we assume that accepting a theory entails being committed either to its literal truth or at least to its empirical adequacy in van Fraassen's sense (1980)—that is, to the theory's being true about what is observable. If accepting a theory entails being committed to the literal truth of its empirical consequences, then accepting an inconsistent theory entails being committed to inconsistent sets of consequences. Thus, if it is irrational to knowingly accept a set of inconsistent sentences as true, then it is irrational to accept the Maxwell–Lorentz scheme if one is aware of its logical structure.

Yet this problem disappears if in accepting a theory, we are committed to something weaker than the truth of the theory's empirical consequences. I want to suggest that in accepting a theory, our commitment is only that the theory allows us to construct successful models of the phenomena in its domain, where part of what it is for a model to be successful is that it represents the phenomenon at issue to whatever degree of accuracy is appropriate in the case at issue. That is, in accepting a theory we are committed to the claim that the theory is *reliable*, but we are not committed to its literal *truth* or even just of its empirical consequences. This does not mean that we have to be instrumentalists. Our commitment might also extend to the ontology or the 'mechanisms' postulated by the theory. Thus, a scientific realist might be committed to the reality of electrons and of the electromagnetic field, yet demand only that electromagnetic models represent the behavior of these 'unobservables' reliably, while an empiricist could be content with the fact that the models are reliable as far as the theory's observable consequences are concerned.

If acceptance involves only a commitment to the reliability of a theory, then accepting an inconsistent theory can be compatible with our standards of rationality, as long as inconsistent consequences of the theory agree approximately and to the appropriate degree of accuracy. Thus, instead of Norton's and Smith's condition that an inconsistent theory must have consistent subsets which capture all the theory's acceptable consequences, I want to propose that our commitment can extend to mutually inconsistent subsets of a theory as long as predictions based on mutually inconsistent subsets agree approximately.

This weaker constraint is in fact satisfied by classical electrodynamics. Given energy conservation, the Lorentz force will do a good job of representing the motion of a charged particle only if the energy of the charge is very large compared with the energy radiated. In that case, the error we make in ignoring the radiation losses implied by energy conservation is negligible. If one plugs in the numbers, it turns out that for an electron, radiative effects would influence the motion of the particle appreciably only for phenomena characterized by times on the order of 10^{-24} sec (such as that of a force that is applied to an electron only for a period of 10^{-24} sec) or by distances on the order of 10^{-13} cm (Jackson 1975, 781–782). These times and lengths lie well outside the theory's empirical limit of validity and within a domain where quantum mechanical effects become important. Within its domain of validity, the theory is approximately consistent: Predictions based on the Maxwell equations and the Lorentz force law, although strictly speaking inconsistent, given energy conservation, agree within any reasonable limit of accuracy.

The flip side of the point that the theory is approximately consistent within a certain domain is that the theory puts limits on its domain of applicability *from within*, as it were. Independently of any empirical considerations, we can know that the theory would not be applicable to phenomena involving very short distances and timescales. For in that case the energy characteristic of the phenomenon is comparable in magnitude with that of the radiation loss, and predictions based on the Lorentz law would appreciably disagree with the requirement of energy conservation. We can contrast this with the case of Newtonian classical mechanics. Today we believe that the (nonrelativistic) theory does not apply to phenomena involving very high speeds or very short distances. But these limits to the theory's domain of applicability had to be discovered empirically, and were not dictated by the theory itself. Unlike standard classical electrodynamics, Newtonian mechanics has no internal limits of reliability. Now, as it turns out, classical electrodynamics becomes empirically inapplicable several orders of magnitude before its internal limit of application is reached. But this does not conflict with the claim that there is such an in-principle limit.

One may also wish to put the point differently: It is precisely because classical electrodynamics is taken to have a limited domain of application that the theory's inconsistency is acceptable. As a candidate for a universal physics, an inconsistent theory would be unacceptable. Yet a theory with a limited domain of validity may be inconsistent as long as the inconsistency does not notably infect predictions within its domain. This is how the physics community by and large appears to view the situation. Before the development of quantum theories the question of the consistency of a classical particle–field theory was of central concern to research in theoretical physics. But with the advent of quantum physics, interest in developing a coherent classical theory seems to have declined rapidly. As Philip Pearle puts it in his review of classical electron models: "The state of the classical theory of the electron theory reminds one of a house under construction that was abandoned by its workmen upon receiving news of an approaching plague. The plague in this case, of course, was quantum theory" (Pearle 1982, 213). And in fact Dirac's attempt to derive a particle equation of motion that is consistent with the Maxwell equations and energy conservation (which we will examine in the next chapter) was to a large extent motivated by the hope of using a consistent classical theory to help in solving problems of the quantum theory.

Thus, an additional constraint on the permissibility of an inconsistent theory is that it cannot be a candidate for a universal physics. Yet again this constraint is significantly weaker than Norton's constraint, according to which inconsistent theories are permissible only as guides to consistent theories.

This brings me to the second disagreement I have with Norton's account. In line with the traditional worry that inconsistent theories allow us to derive arbitrary conclusions, Norton holds that the consequences of inconsistent theories can be of no interest to us, unless approximately the same conclusions can also be derived from a consistent theory. Thus, Norton concludes his discussion of Newtonian cosmology by saying:

> In sum, my proposal is that the content driven control of anarchy can be justified as meta-level arguments designed to arrive at results of an unknown, consistent correction to the inconsistent theory. The preferred conclusions that are picked out are not interesting as inferences within an inconsistent theory, since everything can be inferred there. Rather *they interest us solely in so far as they match or approximate results of the corrected, consistent theory*. (Norton 2002, 194; my emphasis)

Inconsistent theories, according to Norton's view, can play a certain heuristic role, but cannot on their own provide us with reasons for accepting any of their consequences. Thus, the inconsistency of Newtonian cosmology, according to Norton, eventually served as a guide to the discovery of a consistent relativistic theory of gravitation, just as the old quantum theory of blackbody radiation served as a heuristic guide in the development of quantum mechanics. Even though Norton allows inconsistent theories to play an important role in the process of scientific theorizing, he seems to agree with traditional worries about inconsistency in one important respect. Like the traditional view, Norton does not believe that the *best* theory in a certain domain and an *end product* of scientific theorizing could turn out to be inconsistent.

Among those who do not want to reject inconsistent theories outright, this view appears to be widespread. Thus Smith argues that considering inconsistent "proto-theories" can turn out to be scientifically fruitful: "Indeed, the use of inconsistent representations of the world as heuristic guideposts to consistent theories is an important part of scientific discovery and deserves more philosophical attention" (Smith 1988, 429). According to Smith's interpretation of his case studies, the inconsistent principles to which scientists appeal ought to be understood as projections of fragments of yet-to-be-discovered consistent theories in the domain at issue. Philip Kitcher (2002a) appears to hold a similar view, as we have seen in chapter 1.[8]

Classical electrodynamics, however, is not a preliminary theory in the way the old quantum theory of blackbody radiation might be thought to be. The theory has reached a certain stage of completion and appears to be, in some sense, an end product of physical theorizing. But, one might object, has classical electrodynamics not been replaced by quantum electrodynamics? Thus, has classical electrodynamics not been a stepping-stone in the history of physics, analogous to Norton's examples? This objection, however, glosses over an important distinction. Classical electrodynamics is no longer regarded as the most 'fundamental' theory governing the interaction of charged particles with electromagnetic fields. In this sense, one might say, it has been 'replaced.' Yet it remains the most successful and most appropriate theory for modeling phenomena in its domain. Trying to use quantum electrodynamics to model classical phenomena—that is, phenomena characterized by classical length and energy scales—would be grossly inadequate, if it were possible at all. As far as the modeling of classical phenomena is concerned, quantum electrodynamics has not replaced the classical theory; rather, it has helped to establish limits to the theory's domain of validity and, insofar as the classical theory can be shown to be a limit of the quantum theory, the quantum theory allows us to explain certain salient features of the classical theory.

Nevertheless, in justifying the use of the classical theory purely within its domain, scientists do not need to appeal to a quantum theory. By contrast, the old quantum theory of blackbody radiation is no longer regarded as the best theory for modeling atomic phenomena and has been replaced *in its domain of application* by quantum mechanics. Similarly, Newtonian *cosmology* has been replaced by Einstein's general theory of relativity, despite the fact that Newtonian physics remains the most appropriate theory for the mechanics of medium-sized objects. Classical electrodynamics, unlike the theories usually discussed by philosophers interested in inconsistency, is what Fritz Rohrlich calls an *established theory*—that is, a theory with known validity limits that coheres well with other theories and is empirically well supported within its domain (Rohrlich and Hardin 1983; Rohrlich 1988).

Moreover, in contrast to the case of Newtonian cosmology, the considerations that block the derivation of arbitrary conclusions in classical electrodynamics cannot be construed as a guide to a potentially consistent theory—in this case, quantum electrodynamics. If anything, the relationship has been the reverse historically: One of the main motivations for attempts to arrive at a satisfactory and consistent classical theory of point charges was that some of the same problems faced by the classical theory reemerged for quantum electrodynamics. The hope was that a consistent classical theory could function as a guide for constructing a consistent quantum theory. This hope has not been fulfilled.

I will argue in chapter 3 that there is no conceptually unproblematic classical particle equation of motion fully consistent with the Maxwell equations and energy conservation from which the Lorentz force equation could be derived as an approximation. But can we not think of the Maxwell–Lorentz theory as an approximation to an as-yet-undiscovered theory? That is, could we not, in keeping with Norton's suggestion, assume that the consequences of the inconsistent theory are accepted only provisionally, recognizing that if they cannot ultimately be backed up by a "corrected, consistent theory," then they can be of no interest to us and should be discarded?

This suggestion cannot, however, account for the important role of the Maxwell–Lorentz theory (and of particle–field theories more generally) in modern physics. The theory has since the early 1900s become a central part of modern physics, and one or more courses on some amalgam of the particle theory and the continuum theory form an integral part of any physics student's education. The particle theory is extremely successful predictively, even if the goal of developing a fully consistent and conceptually unproblematic theory has proved elusive so far. Thus it seems that the results of classical electrodynamics within its domain of application, similar to those of classical mechanics, are here to stay quite independently of whether or not physics will ever be able to solve the foundational problems posed by particle–field theories. The results of the existing classical theory are of interest to us and, I submit, the classical theory has explanatory power, even if, as appears likely, physicists will never develop a corrected, consistent theory.

Dudley Shapere has argued that "there can be no guarantee that we must always find a consistent reinterpretation of our inconsistent but workable techniques and ideas" (Shapere 1984a, 235). Classical electrodynamics appears to be the paradigm case of a theory where a consistent "reinterpretation" will not be forthcoming. But, as Shapere points out, it would be a mistake, therefore, to reject such theories: "We may have to live with them, learning in the process that, whatever its advantages, consistency cannot be a requirement which we impose on our ideas and techniques on pain of their rejection if they fail to satisfy it" (Shapere 1984a, 235–236).

5. Conclusion

I have argued that the main approach to modeling the interactions between classical microscopic particles and electromagnetic fields is inconsistent. If, as I will argue in the next chapter, we cannot think of this approach as an approximation to some other conceptually unproblematic and fully consistent theory covering the same phenomena, the existence of this inconsistent theoretical scheme has profound consequences for the ways we ought to think about scientific theorizing. Contrary to what many philosophers still seem to consider 'philosophical common sense,' a theoretical scheme can be inconsistent and yet be successful. What is more, inconsistencies can play a role far beyond that of being a provisional guide to the development of consistent successors. Even without the certain prospect of a 'correct' theory waiting in the wings, very good and interesting physics can be done with an inconsistent theory. Finally, the commitment of physicists to such theories need not be restricted to a single consistent subset of the theory's consequences. How is this possible? I have argued that the worry about inconsistent theories can be attributed at least in part to what I take to be a mistaken view on theory acceptance. If we replace a commitment to the literal truth of a theory's empirical consequences with a commitment to a theory's reliability, then content-driven constraints on permissible derivations can ensure that accepting an inconsistent theory need not violate our standards of rationality.

The idea that theory acceptance involves only a commitment to the approximate fit of the models constructed with the help of the theory fits well with the recent 'model-based' accounts of scientific theories, which I mentioned in the introduction. These accounts stress that representational models are usually derived with the help of a host of assumptions in addition to a theory's laws; indeed, inconsistent theories need to include additional 'metaconstraints' restricting the derivations allowed by the theory. 'Model-based' accounts of scientific reasoning are suggestive of older instrumentalist accounts of scientific theories. But once we allow theories to have a richer interpretive framework than only a mapping function fixing the ontology, it is possible that even an inconsistent theory can offer a picture of the world that goes beyond merely providing a disunified set of instrumentally successful models of individual phenomena. According to the theory there are electromagnetic fields, which interact with charged particles in two ways. Fields carry energy and momentum, and the interactions between electric charges and fields satisfy various locality principles. The theory provides us with a contentful account of 'what the world is like' without, however, delineating a coherent class of physically possible worlds, since the mathematical tools available for making this account precise are inconsistent. And just as physicists can learn what the theory says about the world without worrying (too much) about the consistency of the theory, so it is possible to philosophically investigate many aspects of the conceptual structure of the theory in interesting and fruitful ways, as Marc Lange (2002), for example, has done, without ever mentioning problems concerning the theory's consistency.

3

In Search of Coherence

1. Introduction

In the previous chapter I argued that the standard theoretical scheme for modeling classical electromagnetic phenomena involving the interaction of charges and fields is inconsistent. This scheme treats charged particles as point particles and takes particle–field systems to be governed by the Maxwell–Lorentz equations and the principle of energy–momentum conservation. My claim that *this* scheme is inconsistent ought to be uncontroversial, even if it may come as a surprise to many who have some familiarity with standard textbook presentations of the theory. Yet I want to claim something more: namely, that there is no fully consistent and conceptually unproblematic theory for modeling the motion of charged particles in electromagnetic fields that is more successful than the scheme discussed in chapter 2. And I suspect that this stronger thesis will be much more controversial—not least because the very fact that the standard scheme is inconsistent probably constitutes a powerful reason for seeing this scheme as a mere approximation to a consistent classical theory. In this chapter I will argue for the stronger claim by examining several candidates for a 'fundamental' classical theory to which the standard approach might be an approximation.

In section 2, I will discuss the view that particle theories ought to be understood as approximations to a theory of continuous charge distributions. I will argue that the main problem with taking this theory to be the fundamental theory of the interaction of classical charges and fields is that it is in an important sense incomplete. Without substantive additional assumptions concerning how charged particles are to be modeled, the theory cannot be understood as describing the behavior of particle–field systems in principle. In addition I will show that even considered on its own, the theory of continuous charge distributions poses problems for the view that theories ought to delineate a coherent class of possible worlds.

In section 3, I discuss various alternative equations of motion for charged particles that include a *radiation reaction term*. The upshot of this discussion will

48 *Inconsistency, Asymmetry, and Non-Locality*

be that there is no conceptually unproblematic equation of motion that is fully consistent with the Maxwell equations and energy–momentum conservation.

In section 4, I argue that the ways in which various equations of motion are constructed show that it may not always be possible to distinguish clearly, in scientific theorizing, a foundational project of arriving at a set of 'exact' fundamental equations delineating the possible worlds allowed by a theory and a pragmatic project of applying these equations, which is guided by considerations of simplicity and mathematical tractability. As we will see, full consistency is not an overriding concern in attempts to arrive at a fundamental dynamics for classical charged particles.

2. Charged Dusts

2.1. *The Consistency Proof*

My claim that the standard approach to classical electrodynamics is inconsistent may be surprising even to many who have studied the theory in some detail. For there is a standard derivation of the principle of energy conservation that can be found in textbooks on classical electrodynamics that seems to show that the theory *is* consistent. Here is how this consistency proof goes.

From the Maxwell equations together with the Lorentz force law one can derive an equality which expresses a principle of energy conservation just in case the field energy is identified with the standard expression. For it follows from the Maxwell equations that

$$\int_V \frac{1}{4\pi c} \nabla \cdot (\mathbf{E} \times \mathbf{B}) d^3x = -\frac{\partial}{\partial t} \int_V \frac{\mathbf{E} \cdot \mathbf{E} + \mathbf{B} \cdot \mathbf{B}}{8\pi} d^3x - \int_V \mathbf{J} \cdot \mathbf{E} d^3x. \quad (3.1)$$

The term on the left-hand side, according to the definition and interpretation of the Poynting vector, gives the energy flowing out of the volume V. The first term on the right is the change of the field energy and the second term, according to the Lorentz law, gives the change of the energy of the charges. Thus,

$$\oint \mathbf{S} \cdot da = -\frac{\partial}{\partial t} \int_V u d^3x - \int_V \mathbf{J} \cdot \mathbf{E} \, d^3x = -\frac{d}{dt}(E_{field} + E_{charges}). \quad (3.2)$$

That is, the energy flow out of the volume V is exactly balanced by the sum of the energy of the field and that of the charges contained in the volume. And this equality that expresses energy conservation is derived from the Maxwell equations in conjunction with the Lorentz force law.

This derivation closely resembles the argument I gave above to show that the theory of point charges is *inconsistent*, except that my argument crucially involved a self-energy term (see equation (2.14)), which is missing in this derivation. So something must have gone wrong here. Is the consistency proof found in standard textbooks mistaken? What has gone wrong is that we were not sufficiently careful about the ontology of the theory. The consistency proof goes through only for a theory of *continuous charge distributions*. In that case the principle of energy conservation is satisfied exactly, since for infinitesimal 'particles' there is no contribution from the self-fields to the equation of motion, because the field associated with such a particle is likewise infinitesimal. That is, the work done by the *external* fields on an infinitesimal charge is equal to the work done by the *total* fields. Equation (2.14) expresses the energy balance correctly, but in the case of infinitesimal fields (2.14) is equivalent to (3.1) and (3.2).

Another way to make this point is that the Lorentz force is proportional to the charge q, while the energy radiated by a charge (and, hence, the 'self-force') is proportional to q^2. An infinitesimal 'particle' can be understood as the result of taking the limit $m \to 0$ for a finitely charged point particle with mass m and charge q while keeping the ratio q/m constant. Now, according to Newton's second law, the Lorentz force contributes on the order of q/m to the acceleration, while the radiation reaction force is of order q^2/m, and hence contributes nothing in the limit of infinitesimal charges. Infinitesimal particles experience no self-force.

Thus, the Maxwell–Lorentz theory of continuous infinitesimal charge distributions—or *charged dusts*—is consistent, while the theory of finitely charged point particles with the Lorentz force equation of motion is not. In fact, there is a rather tight connection between the three main principles—the Maxwell equations, the Lorentz force, and the expression for the field energy—in the theory of charged dusts. One can assume the Maxwell–Lorentz equations and energy conservation, and then derive the energy expression, along the lines I have suggested. Alternatively one can begin with the Maxwell equations and the expression for the field energy to derive the Lorentz force.

It is striking how difficult it is to detect this crucial difference between the particle theory and the continuum theory in standard textbook discussions of the principle of energy conservation. In fact, one finds the more or less explicit suggestion that the consistency proof for the continuous distribution theory unproblematically carries over to a theory of localized charged particles as well. Thus Landau and Lifshitz in their classic text on field theory (1951, 75–76) derive the principle of energy conservation for continuous charge distributions

$$\int_V \mathbf{J} \cdot \mathbf{E} d^3x + \int_V \frac{\partial u}{\partial t} d^3x + \oint \mathbf{S} \cdot d\mathbf{a} = 0, \tag{3.3}$$

and without further comment write the first integral on the left for discrete (point) charges as $\sum e\mathbf{v} \cdot \mathbf{E}$. They then claim that according to the Lorentz force equation of motion, this expression is equal to the change in the kinetic energy of the charges

$$\sum e\mathbf{v} \cdot \mathbf{E} = \frac{d}{dt} E_{kin}, \tag{3.4}$$

which results in a principle of energy conservation for a system of charged particles. Yet in fact, for a single point charge, the integral over the current ought to be

$$\int_V \mathbf{J} \cdot \mathbf{E} d^3x = e\mathbf{v} \cdot \mathbf{E}_{ext} + e\mathbf{v} \cdot \mathbf{E}_{self}, \tag{3.5}$$

including a contribution from the self-fields, which Landau and Lifschitz simply ignore.

Jackson presents essentially the same derivation as Landau and Lifschitz, explicitly introducing it as a derivation applying to continuous distributions of charge and current (Jackson 1999). Curiously, however, the derivation occurs in a section titled "Conservation of Energy and Momentum for a System of Charged *Particles* and Electromagnetic Fields" (236; my emphasis), and Jackson states that "for a single charge q the rate of doing work by external fields **E** and **B** is q**v**·**E**" (236), before introducing the equivalent expression for continuous charge distributions.[1] Thus, the subtle but important distinction between the two types of systems is likely to be lost on the reader. The entire section is taken up with the derivations of energy and momentum conservation for continuous charge distributions, but this restriction is not made explicit, and there is no discussion of the question of whether or not the derivations apply to charged particles. To be sure, Jackson does explicitly discuss the foundational problems of a theory of charged particles, but not until the very last chapter of his book, where he says that "a completely satisfactory treatment of the reactive effects of radiation does not exist" (781).

2.2. The Principle of Energy Conservation

The consistency proof of the continuum theory can also be understood as an argument for adopting the standard expression for the field energy and the energy flow. If we assume the standard expressions for the energy, then energy is conserved for systems of continuous charge distribution and electromagnetic fields. But why, one might ask, should we adopt the same expressions in the inconsistent particle theory?

Even in the inconsistent theory, the standard expression for the field energy appears to be motivated partly by appealing to considerations of energy–momentum conservation. One argument points to the fact that in regions where there are no charges present, the usual expressions for the energy and momentum associated with the electromagnetic field ensure that field energy and momentum are conserved. From the source-free Maxwell equations one can derive (similar to the derivation above):

$$\int_V \left[\frac{\partial}{8\pi \partial t}(\mathbf{E} \cdot \mathbf{E} + \mathbf{B} \cdot \mathbf{B}) + \frac{1}{4\pi c} \nabla \cdot (\mathbf{E} \times \mathbf{B}) \right] d^3x = 0. \quad (3.6)$$

This equation expresses the conservation of energy if we identify the energy density of the field as $u = (8\pi)^{-1}(\mathbf{E} \cdot \mathbf{E} + \mathbf{B} \cdot \mathbf{B})$ and the energy flow with the *Poynting vector* $\mathbf{S} = c(4\pi)^{-1}(\mathbf{E} \times \mathbf{B})$, as defined above. For it then follows from Gauss's Law that

$$\int_V \frac{\partial u}{\partial t} d^3x + \oint \mathbf{S} \cdot d\mathbf{a} = 0. \quad (3.7)$$

That is, the change in field energy within the volume is equal to the net amount of energy flowing out of the volume. The relativistic generalization of this expression is the requirement that the four-divergence of the *energy–momentum tensor* vanishes: $\partial_\alpha T^{\alpha\beta} = 0$ (with $T^{00} = u$ and $T^{0i} = c^{-1}S_i$).

Considering charge-free regions does not, of course, tell us how energy is transferred from charged particles to the field and vice versa. Discussions of energy conservation for continuous charge distributions in textbooks suggest another strategy to motivate the standard expression for the field energy (even though this strategy is not made explicit). One can derive the standard energy tensor from the requirement that the theory of continuous charge distributions should satisfy the principle of energy conservation, and then carry this expression for the field energy over to the theory of finitely charged point particles.[2] That is, on this view, considerations of consistency lead to the adoption of the standard energy tensor in the theory of continuous charge distributions, which then, for reasons of continuity between the different theories, is adopted in the inconsistent particle theory as well.

Finally, there is empirical support for the standard expression. Specifically, there seems to be good empirical evidence for the claim that the power radiated by an accelerating charge is proportional to γ^4 (where $\gamma = (1-v/c)^{-1/2}$), which follows if the energy flow is given by the Poynting vector. This alone does not uniquely fix an expression for the field energy, but it does exclude expressions for the field energy that are consistent with the Lorentz equation of motion for particle theories. While it is possible to define the energy of the electromagnetic field in a way that is consistent with the Lorentz equation of motion,[3] the problem with such a definition is (as one would expect) that the resulting expression differs from the standard definition by just the right amount to cancel the radiation term. Thus, this alternative expression for the field energy implies that accelerated charges do not radiate energy, contrary to what we observe empirically.

2.3. Discrete Particles as Approximations to Continuous Distributions

As I have suggested in the previous two chapters, one proposal for rendering the inconsistency of the particle scheme relatively benign and reducing the threat to traditional conceptions of scientific theories might be to maintain that discrete particles should be understood as mathematical idealizations within the context of a consistent theory of continuous charge distributions. From a foundational standpoint such a view seems preferable. The problem with the continuum theory, however, is that on its own it is at best of very limited empirical use. There is much evidence to suggest that many microscopic electromagnetic phenomena ought to be modeled as involving discrete charged particles, and as soon as we introduce compact localizations of charges (in the form either of spatially extended charged particles or of finitely charged point particles), self-energy effects, which are absent in a continuum theory, become important.

A historically particularly interesting example of the failure of continuum electrodynamics to represent 'particle phenomena' adequately is the case of synchrotron radiation, which I discussed briefly in chapter 2. At first physicists did in fact represent the stream of electrons orbiting in a synchrotron accelerator as continuous, constant current. Since it follows from the Maxwell equations that static current distributions, which satisfy the condition $\partial j^\mu(x)/\partial t = 0$, do not radiate (see Pearle 1982, sec. 6), physicists did not expect any radiation to be associated with synchrotron charges. Thus, it came as a surprise when the radiation, which happens to occur in the visible part of the electromagnetic spectrum, was discovered purely by accident; and the discovery was taken to show that electrons in a synchrotron have to be modeled as circularly accelerating discrete particles rather than as a continuous distribution.

Now, one might think that we should be able to model discrete charged particles within a continuum theory by differences in the charge density of a continuous distribution, and then simply apply the ordinary Lorentz law to the infinitesimal 'point charges' of the distribution. But if the only force acting on a continuous distribution is the Lorentz force, then we would expect that local regions of higher charge density corresponding to charged particles are not stable and flow apart, due to Coulomb repulsions among the different parts of the region. That is, discrete, extended particles that retain their integrity through time are inconsistent with the Lorentz force as the only force acting on the particles. The continuum theory on its own is inconsistent with the existence of the particle phenomena it is assumed to cover.

An obvious response to this worry is to introduce additional (perhaps not further specified) cohesive forces that can ensure the stability of regions of higher charge density. Yet once we introduce such forces, the Lorentz law alone no longer correctly predicts the behavior of such regions as a whole and we are led to a picture with discrete, finitely charged particles with finite radiation effects affecting the motion of the center of mass of such a particle. Thus, one needs to distinguish carefully between a theory of continuous charge distributions and one of extended, yet discrete, charged particles. While in the former case the Lorentz force law describes the motion of dust particles consistent with energy conservation, in the latter case, taking the center of mass motion for a charged particle to be given by the Lorentz force law is inconsistent with energy conservation.

The idea that the application of laws frequently involves various forms of approximations such that the equations characterizing the resulting representational models are, strictly speaking, inconsistent with the laws from which they are derived is quite familiar. To mention a simple example, in deriving the pendulum law in classical mechanics from Newton's equations of motion, one customarily makes the small angle approximation which leads to an equation of motion that, strictly speaking, is inconsistent with Newton's laws, since the latter tell us that the force on the bob of the pendulum is proportional to the sine of the angle and not to the angle. Yet one can plausibly maintain that the pendulum is governed by some exact (if perhaps rather complicated) equation of motion that is in principle derivable from the fundamental theory. That is, we can plausibly assume that Newton's laws (including the law of gravitation) on their own govern the behavior of a pendulum in the sense that there is a model-theoretic model of the laws that accurately represents the behavior of the pendulum. Thus, there is nothing particularly problematic about the fact that the harmonic oscillator pendulum equation is, strictly speaking, inconsistent with the dynamical laws from which it is derived.

The theory of continuous charge distributions, however, cannot function as a fundamental theory of phenomena involving charged particles in the same way. For the laws of this theory have *no* model-theoretic models that can (even in principle) adequately represent the behavior of charged particles, since they do not on their own allow for any localized electromagnetic objects that retain their integrity through time. Thus, the theory in an important sense is incomplete, if it is understood as a theory of charged particles as well; it is incomplete without further assumptions intended to ensure the stability of compact localizations of charges. Unlike in the case of the pendulum, there is no fully consistent and complete classical theory waiting in the wings, as it were, which we could understand as governing the phenomena in principle. Rather than constituting an approximation to the continuum theory, the particle scheme provides an *extension* of the continuum theory into a domain not governed by the theory of continuous distributions alone. And as we shall see in section 3, there is no fully consistent alternative extension of the continuum theory to which the Maxwell–Lorentz theory could be seen as an approximation in the proper sense. Thus, insisting that the continuum theory is the basic theory does not render the inconsistency of the particle theory unproblematic. On that conception the puzzle is how a consistent theory can be inconsistently extended to cover phenomena of which it has no models. This puzzle is every bit as troublesome for traditional views of theories as is the puzzle of how an inconsistent theory can be successful.

2.4. Shell Crossing

Before turning to a discussion of alternative equations of motion for discrete charged particles, I want to show that, even considered on its own, the theory of continuous distributions is not without problems. For there is a wide class of intuitively reasonable initial conditions for which the Maxwell–Lorentz equations for continuous charge distributions do not have global solutions. Thus even this seemingly well-behaved theory spells trouble for the view that the laws of a theory delineate classes of physically possible worlds.

For a charged dust one can formulate a pure Cauchy problem for the Maxwell–Lorentz equations (Parrott 1987, secs. 5.1–5.3). More specifically, one can show that initial conditions consisting of the initial fields $\mathbf{E}(t=0, \mathbf{x})$, $\mathbf{B}(t=0, \mathbf{x})$ and the velocity of the dust $\mathbf{v}(t=0, \mathbf{x})$ determine a unique solution to the Maxwell–Lorentz equations on an open subset of R^4, if one assumes that $\mathbf{B}_0 = \mathbf{B}(t=0, \mathbf{x})$ satisfies $\nabla \cdot \mathbf{B}_0 = 0$. (The charge density ρ need not be introduced separately, but is defined via Coulomb's law $\nabla \cdot \mathbf{E} = 4\pi\gamma\rho$, where $\gamma(\mathbf{v})$ is the factor familiar from relativity theory.) Unlike the Maxwell equations alone, the Maxwell–Lorentz equations are only quasi-linear, but nevertheless the theory seems to fit the standard view of theories rather well. The initial conditions can be freely chosen, with one constraint, and the laws then determine how a 'world' with a given initial state has to evolve. In other words, the Maxwell–Lorentz equations seem to delineate the class of electrodynamically possible worlds, where different worlds are picked out by different sets of contingent initial conditions.

The problem, however, is that standard existence and uniqueness proofs ensure only that there are *local* solutions to the Maxwell–Lorentz equations, defined on some open set $t_1 < t < t_2$, $r_1 < r < r_2$. And there is in fact a large class of seemingly physically reasonable initial conditions for which *global* solutions to the equations can easily be shown not to exist: namely, spherically symmetric charge distributions with zero initial velocity for which the dust density thins out sufficiently rapidly as $r \to \infty$.

Each dust particle in this case will experience an acceleration proportional to Q/r^2, where Q is the total charge inside the sphere with radius r (centered at the origin) and is constant on the worldline of the particle. Since the initial velocity is zero, dust particles will move outward for $t > 0$ with ever increasing velocity and $r(\infty) = \infty$. Q/r functions as the potential energy of a dust particle at distance r from the origin. If a particle has a higher potential energy at $t=0$, then it will end up with a higher terminal velocity $v(\infty)$. Now consider two dust particles, 1 and 2, with $r_1(0) < r_2(0)$, such that the inner particle has a higher potential energy; that is, we assume that $Q_1/r_1 > Q_2/r_2$. This will be the case whenever the charge density thins out sufficiently rapidly with r, so that the difference in the total charges contained in the spheres with radii r_1 and r_2 is more than made up for by the increase in radius. Intuitively, the inner particle will have greater potential energy if the total charge contained in the sphere increases only slightly as the radius of the sphere increases appreciably.

Since particle 1 starts inside the sphere of particle 2, but its terminal velocity $v_1(\infty)$ is larger than that of particle 2, there must be a time t^* at which $r_1(t^*) = r_2(t^*)$. That is, if we take two particles with the same angular coordinates, the worldlines of the particles intersect, which is impossible. For if two worldlines of infinitesimal dust particles were to intersect, the velocity of the dust would not be well defined at that point, since two different velocity vectors would be associated with one and the same point. One might think that this problem ought to be avoidable by some mathematical trick which lets different dust particles pass through each other. I am not sure that this can be done, since speaking of 'dust particles' is just an intuitive way of speaking of points in a continuous distribution. But even if this problem could be solved, the fact that particle 1 ends up outside the sphere on which particle 2 is located, even though it started inside it, is inconsistent with Q being constant on the worldline of a charge, which is a consequence of the Maxwell equations. That is, models with shell crossing are inconsistent with the Maxwell equations.

The upshot of this discussion is that there are no global solutions to the Maxwell–Lorentz equations for the kind of case we discussed; the solutions cannot be extended beyond some time $t^* > 0$, and (since the equations are time-symmetric) backward in time beyond a time $-t^* < 0$. Moreover, it seems plausible that the problem is not limited to the specific case discussed, and that intersecting worldlines are a typical feature characterizing many more types of initial conditions than the ones we discussed.

What are we to conclude from this? One possible reaction might be to argue that the case shows that initial conditions cannot be chosen as freely as we might initially have thought.[4] This is a point familiar from discussions of closed causal loops in time travel scenarios. There it is argued that the impossibility of engineering the death of one's own grandfather before the conception of one's father, for example, does not show that causal loops are impossible, but only that the relevant initial conditions could not possibly obtain. Applied to our case, one might argue that what appear to be physically reasonable initial conditions are shown to be physically impossible after all. But in the current context this response is not very plausible. Unlike the case of closed causal loops, it is not clear how the breakdown of the solution in the future could have any effects, as it were, on an experimenter's effort to set up a spherically symmetric charge distribution which thins out relatively rapidly.

Parrott suggests that the possibility of intersecting worldlines calls into doubt whether the retardation condition is consistent with the Maxwell–Lorentz equations, since, as he says, retarded solutions must be defined on the entire backward light cones of each field point (Parrott 1987, 187). The worry seems to be this. According to the retarded field representation, the field at each point depends on what happens on the entire backward light cone erected at that point. So if there is a finite time $-t^* < 0$ beyond which the solution cannot be extended, then the retarded solution is not well defined. But the problem seems to lie more directly with the nonexistence of global solutions. For if we set up a *local* initial-value problem for a finite region of space-time, the retarded solution needs to be defined only within that region, since the contribution of charges in the past of the region will be encoded in the incoming field on the initial-value surface. Thus, if we rest content with only local solutions of the Maxwell–Lorentz equations, there is no problem with the retardation condition.

3. Theories with Radiative Reaction

3.1. Abraham's and Lorentz's Models of Extended Charges

In the last section I discussed the theory of continuous charge distributions as a consistent alternative to the standard point particle approach. Aside from the problem concerning the nonexistence of global solutions to the theory, the difficulty with understanding this theory as the fundamental classical field theory is that it does not govern phenomena involving charged particles. Once one introduces compact localizations of charges, radiation effects become important and the (center of mass) motion of such localized finitely charged particles can no longer consistently be described by the Lorentz force equation of motion. In this section I will discuss attempts to try to arrive at a consistent theory of finitely charged particles by adding a radiation term to the particle equation of motion, taking into account the self-fields generated by the charge.

An obvious reason for ignoring the self-field is that the standard theory treats charges as point particles, and the self-field of a point particle with finite charge is infinite at the location of the particle. This can be easily seen by considering the q/r^2 dependence of the Coulomb field associated with a point charge. If the charge q is finite, then the Coulomb field diverges as the radius $r \to 0$. There are two types of responses to the problem of infinities, the first of which is to treat charged particles as extended objects. In that case no infinities arise, since the total finite charge of a particle is smeared out over a finite volume and, hence, is infinitesimal at each point within the particle. The second response to the problem created by the infinite self-fields is to stick with a point particle ontology but try to find some procedure for systematically subtracting part of the infinite self-fields.

Historically, theories of microscopic charged particles were first developed as theories of extended charges. The main precursor to contemporary treatments of classical electrodynamics is Lorentz's theory of the electron, which Lorentz intended to provide a microscopic foundation for nineteenth-century Maxwellian macroscopic electrodynamics by combining an atomic conception of matter with a field-theoretic treatment of electromagnetic phenomena (see Lorentz 1916). In contrast to Maxwellian electrodynamics, which did not clearly distinguish charged matter from electromagnetic fields, Lorentz's theory posited two independently existing and equally fundamental types of quantities—microscopic electric charges and an all-pervasive ether—as the seat of the electromagnetic field.

While Lorentz and others—such as Max Abraham, the other main proponent of a classical theory of the electron—used point particle approximations to calculate the field of an electron in regions sufficiently far from the charge, they maintained that in light of the problem of infinities, an electron could not fundamentally be construed as a point particle. The models for the electron that Lorentz and Abraham considered instead assumed that the electron was a sphere with a uniformly distributed surface charge. The obvious problem with any such model is that since the field associated with one part of an extended charged particle results in repulsive forces on all other parts of the charge, an extended charged particle should be unstable and blow apart.

Abraham tried to solve the stability problem by simply postulating that electrons are perfectly rigid objects, and argued that, in analogy to constraints on rigid bodies in classical mechanics, the constraints that keep the electron rigid do no work and need not be explicitly introduced into an equation of motion in the form of additional nonelectromagnetic forces (see Abraham 1908, 130). One worry one might have about Abraham's proposal is that his solution appears to be avoiding the stability problem rather than solving it. A second worry is that the model is not relativistic, since Abraham's electron always is spherical in a preferred reference frame—the ether rest frame. A variant of this criticism was made by Lorentz, who objected to Abraham's electron model on the grounds that Abraham's theory could not account for the null results in ether drift experiments, such as the Michelson–Morley experiments.

In contrast with Abraham's rigid electron, Lorentz assumed that electrons moving with respect to the ether rest frame contract in accordance with the Lorentz–Fitzgerald contraction hypothesis. That is, for Lorentz, electrons are deformable and are flattened into ellipsoids when they have a nonzero velocity in the ether rest frame. Contrary to Einstein's theory of relativity, Lorentz took this contraction to be a dynamic and not a kinematic effect. The proper relativistic analogue to Lorentz's electron model is a *relativistically rigid* particle. A relativistically rigid electron always is spherical in its instantaneous rest frame and has the shape of a flattened ellipsoid in inertial frames moving with respect to the particle's instantaneous rest frame. In the case of deformable electrons, the stability problem appears to arise with added force. Poincaré proposed in 1905 that the electron's stability might be ensured by internal nonelectromagnetic cohesive forces. This proposal was adopted, at least tentatively, by Lorentz, even though he argued that

Poincaré's internal stresses can ensure only that an electron moving with constant velocity is stable. An electron undergoing arbitrary motion, Lorentz believed, would be unstable.

Interestingly, Lorentz did not think that the apparent instability of his electron was a sufficient reason for rejecting the model.[5] Instead, he pointed to the success of his model in allowing derivation of the invariance of the Maxwell equations under what has come to be known as *Lorentz transformations*, and said:

> Having got thus far, we may proceed as is often done in theoretical physics. We may remove the scaffolding by means of which the system of equations has been built up, and, without troubling ourselves any more about the theory of electrons and the difficulties amidst which it has landed us, we may postulate the above equations as a concise and, as far as we know, accurate description of the phenomena. (Lorentz 1916, 222–223)

Thus, even though Lorentz thought that the hypothesis of a stable electron was inconsistent with the dynamical laws governing charged particles, he took the hypothesis to be useful. These remarks suggest that Lorentz would have had some sympathies for the claim I am defending here: that even an inconsistent theory can be scientifically successful.[6]

A problem for any attempt to make use of relativistically rigid electrons in order to arrive at a completely consistent classical electrodynamics is that even a relativistically rigid electron is inconsistent with the special theory of relativity. Unlike Abraham's model, relativistically rigid electrons do not single out a privileged reference frame. Yet the constraint that the particle always is spherical in its instantaneous rest frame is incompatible with the demand that a force acting on one end of the particle will take a finite time to propagate through the particle. If a force propagating at a finite velocity acts on one end of a particle, the particle should initially look 'dented,' even in its instantaneous rest frame, until the force has propagated through the entire charge. Thus, if one's aim is a *fully* consistent theory, it is unclear whether a theory of extended particles can be of much help.

3.2. The Delayed Differential-Difference Equation

If one assumes as a particular model for the charge distribution a relativistically rigid charged shell, one can derive an equation of motion for a charged particle from the assumption that in addition to the Lorentz force due to any external fields, the charge is affected by the Lorentz force that different parts of the charge exert on one another. The self-force can be given a Taylor series expansion in powers of the derivative of the velocity of the charge. If one ignores nonlinear terms—terms involving combinations of different orders of derivatives—then the expression for the self-force can be summed and one arrives at the following equation of motion:

$$m\dot{\mathbf{v}} = \mathbf{F}_{ext} + \frac{e^2}{3c\,a^2}\left[\mathbf{v}\left(t - \frac{2a}{c}\right) - \mathbf{v}(t)\right], \tag{3.8}$$

where a is the radius of the charge.

This equation is a *delayed differential-difference equation*, which gives the acceleration at one time in terms of the velocity at that time and at a slightly earlier time. The time delay corresponds to the time it takes for light to travel through the electric charge. Intuitively the presence of the delay factor can be explained by the fact that a point P on the charged shell at a time t should be influenced not by the field produced by the charges on the opposite side of the shell at t, since these fields take a finite time to reach P, but by the fields produced at times slightly earlier than t. (The equation seems to tell us that the effects of fields of nearby points cancel each other out.)

The differential-difference equation of motion has the advantage over the standard Lorentz equation of motion of including a radiation reaction term. Thus the equation seems to get us closer, as it were, to a consistent classical theory. But of course the resulting theory is not *fully* consistent. For, first, the equation is derived from a model for charged particles that is not consistent with special relativity and, second, the equation is derived by ignoring an infinite number of nonlinear terms in the series expansion for the self-force.

Moreover, there are pragmatic reasons for preferring the standard approach. The Lorentz force equation of motion for a point charge is a Newtonian equation of motion. By contrast, the differential-difference equation relates the acceleration at one time to the velocity at some other time, and therefore cannot be solved by merely specifying Newtonian initial conditions of position and velocity at one time. One common approach to solving this type of equation is assuming a solution to the equation in an interval from $t - R/c$ to t and then extending this solution stepwise to all t with the help of the equation. This is equivalent (assuming analyticity—that is, infinite differentiability) to positing a set of infinitely many initial conditions. In general, differential-difference equations are notoriously difficult compared with ordinary differential equations and often are impossible to solve exactly.

The equation also apparently violates classical determinism, which states that a system is deterministic if the state of the system at one time determines its state at all other times. Thus, according to Earman's well-known formulation of Laplacian determinism, a "world W ∈ **W** [where **W** is the collection of all physically possible worlds] is *futuristically* (respectively *historically*) *deterministic* just in case for any W′ ∈ **W**, if W and W′ agree on any time slice [that is, a Cauchy surface], then they agree everywhere" (Earman 1986, 59). Earman explicitly wants to exclude from his definition cases where, as in the case of a differential-difference equation, appeals to analyticity are needed to satisfy the condition. That is, a system is deterministic only if the values of a *finite* set of state-variables at each point determine the state of the system at all other times, and (3.8) does not satisfy this condition.[7]

Some have objected to the approach that leads to the differential-difference equation on the grounds that it needs to rely on specific assumptions about the structure of a charge—assumptions which appear to be untestable. This last worry is the motivation for two alternative attempts at including a radiation reaction term in an equation of motion. The first assumes that microscopic charged particles have no structure and treats charged particles as point particles. The second approach treats charged particles as extended objects but tries to extract an equation of motion that describes their behavior only to the extent that it is structure-independent.

3.3. The Lorentz–Dirac Equation

The point particle theory with radiation reaction term is due to P. A. M. Dirac, who argued for a point particle model of the electron, as opposed to Lorentz's extended particle model, by appealing to the simplicity of the former: "it seems more reasonable to suppose that the electron is too simple a thing for the question of laws governing its structure to arise" (Dirac 1938, 149). His aim, Dirac said, was not to propose a detailed physical model of the structure of the electron, but

> to get a simple scheme of equations which can be used to calculate all the results that can be obtained from experiment. The scheme must be mathematically well-defined and self-consistent, and in agreement with well-established principles, such as the principle of relativity and the conservation of energy momentum. (ibid.)

Schematically, a derivation of Dirac's relativistic equation of motion proceeds by demanding that energy and momentum be conserved in the interaction between the charge and an external field. This demand is imposed by surrounding the space-time trajectory, or worldline, of a charge with a small 'tube' and demanding that the energy change of the charge be balanced by changes in the field energy within the tube and the energy flowing across the surfaces of the tube. That is, roughly, one demands that the difference in the field energy contained in a certain spatial region at two times and the difference in the particle's energy at these times be balanced by the field energy flowing across the boundaries of that region. The equation of motion for the charge is then derived by shrinking the radius of the tube to zero. The difficulty one encounters, however, is that the total energy of the field in the volume is infinite, since it includes the field due to the charge itself, which is infinite at the location of a point charge.

This difficulty can be overcome, however, if we assume that the charged particle is *asymptotically free*—that is, if we assume that there is no force acting on the charge at both past and future infinity. If we take this to imply that the acceleration of the charge vanishes at infinity, the self-field of the charge can be written as consisting of two components: a finite radiation reaction component and an infinite component that has the form of an inertial mass term, since it is multiplied by the acceleration of the charge in the equation of motion. The nonrelativistic equivalent of Dirac's equation of motion, then, has the form

$$m_0 \dot{\mathbf{v}} = \mathbf{F}_{ext} + \left(\frac{2}{3} \frac{e^2}{c^3} \ddot{\mathbf{v}} - m_{em} \dot{\mathbf{v}} \right), \tag{3.9}$$

where the term in parentheses is due to the self-field and the electromagnetic mass term m_{em} is the troublesome infinite term. Dirac proposed that one should treat this term as a contribution to the empirically observable total mass m of the charge, which is finite. This means, however, that one has to postulate an infinite and negative (!) nonelectromagnetic 'bare' mass m_0 of the charge to balance the infinite electromagnetic mass: $m = m_0 + m_{em}$. This on first sight rather startling procedure, which is known as *renormalization*, has become a standard procedure in quantum field theories.

Mathematically, renormalization is a well-defined procedure, since formally the point particle limit is taken only after the two masses are combined. Thus, one never needs to add or subtract infinite quantities. But conceptually the procedure is far from unproblematic. In particular, it is not clear what the status of the diverging negative bare mass is. One problem with having to postulate a negative bare mass is that (contrary to what Rohrlich himself suggests) the theory seems to violate what Rohrlich calls *the principle of the undetectability of small charges* (Rohrlich 1990, 212–213). One demand we place on our theories is that they ought to fit well with, or be appropriately related to, other theories that we accept (which does not, however, mean that theories have to be formally consistent with one another). Thus, Rohrlich argues—and to my mind convincingly—that one desideratum for a charged-particle theory should be that the equation of motion should approach that of a neutral particle in the limit where the charge goes to zero. But if the charge of a particle goes to zero, its electromagnetic mass will go to zero, while its bare mass, which is assumed to be of purely nonelectromagnetic origin, should presumably be unaffected by taking this limit. Thus, the total mass of the particle in the limit of zero charge should be negative and infinite, and the equation of motion does not reduce to a well-defined Newtonian equation of motion.[8]

The renormalized equation of motion is the *Lorentz–Dirac equation*. Its nonrelativistic approximation (which is also known as the *Abraham–Lorentz equation*, since it is the first approximation to an extended-particle equation derived by Lorentz and Abraham) is

$$m\dot{\mathbf{v}} = \mathbf{F}_{ext} + \frac{2e^2}{3c^3}\ddot{\mathbf{v}} = \mathbf{F}_{ext} + m\tau\ddot{\mathbf{v}}, \qquad (3.10)$$

where $\tau = 2e^2/3mc^3$ and m is the renormalized mass. In addition to the Lorentz force $\mathbf{F}_{ext}$ the equation contains a *radiation reaction force* which depends on the derivative of the acceleration of the charge. Equation (3.10) is a third-order differential equation with an additional degree of freedom compared with standard Newtonian equations of motion, which are second-order differential equations. While one needs only to specify initial positions and velocities to solve a Newtonian equation, an additional condition, such as the initial acceleration of the charge, is necessary in the case of the Lorentz–Dirac equation.

The additional degree of freedom is problematic in that it breaks the standard conceptual connection between forces and acceleration. The general solution to (3.10) is

$$m\dot{\mathbf{v}}(t) = e^{t/\tau}\left[m\dot{\mathbf{v}}(0) - \frac{1}{\tau}\int_0^t dt' e^{-t'/\tau}\mathbf{F}(t')\right].[9] \qquad (3.11)$$

For many choices of the initial acceleration this equation allows *runaway* behavior, according to which charges can accelerate continuously, independently of any external field. For example, if the external force is zero, the equation becomes $\dot{\mathbf{v}}(t) = \dot{\mathbf{v}}(0)e^{t/\tau}$, which leads to runaway behavior unless the initial acceleration is zero. Runaway solutions are usually rejected as unphysical, since they allow a charge to accelerate without 'any reason.' These solutions can be avoided if one explicitly imposes the additional constraint that the acceleration of a charge should vanish at future infinity, a constraint that already had to be imposed during the mass renormalization.[10] This constraint is equivalent to demanding that the initial acceleration in (3.11) be given by

$$m\dot{v}(0) = \frac{1}{\tau}\int_0^\infty dt' e^{-t'/\tau}\mathbf{F}(t'). \tag{3.12}$$

The equation of motion then reduces to a second-order integro-differential equation, which requires only Newtonian initial conditions:

$$m\dot{v}(t) = \int_0^\infty ds\, e^{-s}\mathbf{F}(t+\tau s). \tag{3.13}$$

But (3.13) predicts that charges *preaccelerate*—that is, begin to accelerate *before* the onset of any external force. A simple example of this is a charge subject to a brief force pulse modeled by a δ-function. Equation (3.13) predicts that the charge will move with roughly constant velocity until shortly *before* the force is applied, when it will begin to accelerate. After the force pulse, the charge will move with constant velocity. Under the standard interpretation that forces are causes of acceleration, this means that the theory is backward-causal, since charges accelerate as a result of future forces. Moreover, the theory is causally nonlocal in the sense that it allows forces to act where they are not, and even when they are not. I will discuss the issue of non-locality and the role of the asymptotic condition in the theory in more detail in chapter 4.

The main motivation for the Lorentz–Dirac equation is to find an equation of motion for a charged particle that is consistent with energy–momentum conservation. Yet while the theory satisfies energy–momentum conservation *globally*, if the asymptotic condition of vanishing acceleration at both past and future infinity can be imposed, energy is not conserved *locally*. The conservation principle does not hold for an arbitrary space-time volume, but only for volumes that enclose the entire worldline of the charge, and hence the differential, local version of the principle need not hold. In particular, since charges can begin to accelerate before the onset of an external force (radiating energy in the process), changes in the energy of a charge are not balanced locally by external energy supplied to the charge.

There are additional problems for the Lorentz–Dirac equation of motion that concern systems of more than one particle. For one, there are no general existence and uniqueness proofs for solutions in the case of such systems, and thus, it is not clear whether the theory can be extended to cover more than one particle. Moreover, one of the few known two-particle solutions predicts that two oppositely (!) charged particles which initially approach one another head-on do not collide, but instead eventually turn around and move in opposite directions with ever increasing kinetic energies (see Parrott 1987, sec. 5.5). On the most natural understanding, these solutions violate energy and momentum conservation, since they predict that the two-particle system constitutes an unlimited reservoir of energy. One might argue that this is not a problem for the Lorentz–Dirac approach, since the Lorentz–Dirac equation was only shown to be consistent with energy–momentum conservation, if one assumes the condition of asymptotically vanishing acceleration, and this condition is violated for this two-particle solution. Yet given that the reason for trying to derive an equation of motion that includes a radiation reaction force is the fact that the Lorentz force equation of motion violates energy conservation, the violation in the two-particle case is deeply problematic. Moreover, it is not even clear what the justification for applying the Lorentz–Dirac equation to this case is, since the equation can be derived only if one assumes the asymptotic condition, which these solutions violate.

There are arguments to the effect that energy is conserved in this case, either since the infinite self-energy provides an unlimited energy reservoir, or since the kinetic energy associated with a particle with negative bare mass is negative and, hence, decreases as the particle's velocity increases. But it is not clear that it is legitimate to appeal separately to the two components of the empirically observable renormalized mass in this way. For even if the kinetic energy associated with the negative mass decreases as the particle's velocity increases, this decrease in energy is more than made up for by the increase in energy of the positive contribution to the mass. In fact, the two should balance in just the right way so that the kinetic energy of the particle is given by $1/2 m_{ren} v^2$. These arguments, to my mind, illustrate the dangers that are associated with attempts to subtract away the infinities in the self-fields.

One response to the conceptual problems of the Lorentz–Dirac equation is to argue, in analogy to the case of the Lorentz force law, that the equation is reliable only within a certain domain, where once again the theory itself puts limits on the domain of its acceptability. For if the timescales characterizing a problem are large compared with τ, which is on the order of 10^{-23} sec, then preacceleration effects do not contribute appreciably to the motion of a charged particle and the solutions to the equation are, to any reasonable degree of accuracy, causally 'well-behaved.' The limit at which such effects would become significant lies well beyond that imposed empirically by quantum physics. In fact, within its domain of application, predictions for particle orbits based on the Lorentz–Dirac equation agree with those based on the Lorentz force law. Similarly, in the problematic two-particle solution, the two point charges are predicted to turn around and begin their runaway motion only after the distance separating them has become too small to be governed by classical electrodynamics.[11]

Arthur Yaghjian (1992) argues, in an argument endorsed by Rohrlich (1997), that careful attention to the smoothness conditions which the mathematical functions involved have to satisfy can help to avoid the problem of preacceleration. He points out that the derivation of the Lorentz–Dirac equation requires that the functions involved in the equation can be expanded in a Taylor series and, hence, need to be analytic. He then invokes the analyticity condition in two different ways. First, he argues that the analyticity condition is violated by standard toy models to which the theory is applied, such as a charge experiencing a step-function force. If one smoothes out the force function by adding a suitable correction function, then, Yaghjian argues, the charge will not begin to accelerate *before* the onset of any force.[12] Yaghjian's amended equation of motion does not, however, do away with the fact that after the initial onset, the acceleration at one time depends on the force at all future times, as predicted by (3.13). Yaghjian tries to address this problem by appealing to the analyticity condition a second time. He points out that since the force can be expanded in a Taylor series, the acceleration at one time can be expressed in terms of the force at that time and *all the derivatives of the force at that time*. This observation is meant to be sufficient to show that the theory is not conceptually problematic.

But clearly this argument is inadequate as a solution to the problem of pre-acceleration. The dependence of the acceleration at one time on future forces at future times is problematic only if this is a *causal* dependence. There is nothing bothersome about the notion that there is a mere *functional* dependence of the present on the future. But the fact that future forces (and, hence, the present acceleration) may already be functionally determined by present data (i.e., the present values of all the derivatives of the force) in no way undermines the causal claim that the future is causally responsible for the past. Thus, Yaghjian's solution is either unnecessary or unsuccessful. If we do not want to interpret the Lorentz–Dirac equation causally, the backward determination of the equation is unproblematic; but if we do wish to interpret the equation causally, appealing to the Taylor expansion does nothing to alter the causal structure.[13]

3.4. *The Regularized Equation of Motion*

Above, I said that treating charged particles as extended, results in a structure-dependent equation of motion, which has led some physicists, like Dirac, to argue for a point particle approach. But conversely, the 'pathological' behavior of solutions to the Lorentz–Dirac equation for point particles has led physicists to argue that the conception of a (classical) point particle is inherently problematic and that charged particles should be treated as extended objects in classical electrodynamics. Thus Rohrlich, whose 1965 work provides a classic discussion of the point particle approach, claims in a more recent paper that

> in a strict sense the notion of a "classical point charge" is an oxymoron because "classical" and "point" contradict one another: classical physics ceases to be valid at sizes at or below [the quantum mechanical limit of] a Compton wavelength and thus cannot possibly be valid for a point object. (Rohrlich 1997, 1051)

Rohrlich does not distinguish clearly here between the in-principle or internal limits to the theory's applicability and the empirically established limit imposed by quantum phenomena. Yet these are two different things. For even though classical mechanics also becomes empirically inadequate at the Compton wavelength, there appear to be no "contradictions" limiting the domain of applicability of that theory from within. The fact that a theory *for empirical reasons* has a limited domain of application does not imply that trying to apply the theory outside of its domain results in a logical contradiction. By contrast, the Lorentz–Dirac theory can quite naturally be understood to have a built-in limitation on its domain of application due to the fact that the theory's problematic conceptual structure plays a role only for small length and time scales. But this internal limit to the theory's applicability kicks in several magnitudes below the empirical limit given by the Compton wavelength. Nevertheless, Rohrlich's remarks express the relatively common view that the point particle limit of classical electrodynamics ought to be regarded only as an idealization.

Rohrlich (1997) takes the fact that classical treatments of electromagnetic phenomena have a limited domain of applicability to support both the differential-difference equation for extended charges and the pragmatic usefulness of the Lorentz–Dirac equation despite its conceptual problems. Yet there is a third approach (a version of which Rohrlich has adopted most recently) which agrees with Rohrlich's 1997 work that finitely charged point particles ought to be regarded as, strictly speaking, unphysical and as representing only an idealized limit of what must be thought of physically as extended (yet localized) charge distributions, but which does not make any specific assumptions about the structure of a charged particle.

This approach begins with the Lorentz–Dirac equation, but instead of taking it as an exact equation of motion for point particles, treats it as an approximate equation for extended charges. If we ignore radiation effects completely, the motion of charged particles is given by the Lorentz equation of motion. The radiation reaction term of the Lorentz–Dirac equation, which is structure-independent, presents a first correction to the Lorentz equation. Higher-order corrections would depend on the precise structure of charged particles. Even without knowing the precise form of such higher-order corrections, one can show that these terms will be negligible as long as the time during which the acceleration of the charged particle changes appreciably is large enough. More specifically, the timescale over which the acceleration changes appreciably must be large compared with the time it takes light to travel across the charge distribution. That is, the equation of motion of a charged particle is assumed to be of the form

$$\dot{\mathbf{v}} = \frac{1}{m}\mathbf{F}_{ext} + \tau\ddot{\mathbf{v}} + O(\tau\ddot{\mathbf{v}}a). \tag{3.14}$$

The first term on the right is the Lorentz force, and the second term is the (nonrelativistic approximation to the) Lorentz–Dirac radiative reaction term. The higher-order structure-dependent corrections, which depend on the radius a of the charge, are assumed to be negligible.

Taking the Lorentz–Dirac equation as an approximation provides one with an argument for replacing this conceptually problematic equation with one that exhibits neither preacceleration nor runaway behavior through an additional approximation procedure. This procedure — known as *regularization* — reduces the order of the equation of motion from third-order to second-order and consists in replacing the Lorentz–Dirac equation with a second-order differential equation which differs from the third-order Lorentz–Dirac equation only by terms of at most the same order of magnitude as the higher-order structure-dependent corrections (Flanagan and Wald 1996; Quinn and Wald 1996; Poisson 1999).[14] That is, the argument for the regularized equation of motion is that, from the perspective of an ontology of extended charges, the regularized equation is an acceptable approximation wherever the Lorentz–Dirac equation is. Moreover, the regularized equation has the tremendous advantage that as a second-order equation it behaves just like a Newtonian equation of motion and does not predict runaway or preacceleration behavior. Since preacceleration effects from a short force pulse are nonnegligible only immediately before the force is applied, it is argued that acceleration changes over such short timescales lie outside of the domain within which the Lorentz–Dirac equation provides an acceptable approximation.

The regularized equation of motion is derived as follows. We begin by assuming that charged particles ought to be modeled as localized *extended* charge distributions of arbitrary shape. While this does not allow us to derive an exact equation of motion, some very general assumptions about the charge distribution are enough to derive the Lorentz–Dirac equation as a first-order correction to the Lorentz force law, including the radiation reaction term that depends on the derivative of the acceleration, and the order of magnitude of higher-order corrections. Then the third-order Lorentz–Dirac equation is replaced by a second-order equation which agrees with the Lorentz–Dirac equation up to the order of magnitude of the structure-dependent higher-order corrections. This is done by differentiating the equation once and then plugging in the result for $\ddot{v}$ on the right side of the original equation, dropping all higher-order terms. This leads to the second-order regularized equation of motion

$$\dot{v} = \frac{1}{m}\mathbf{F}_{ext} + \frac{\tau}{m}\dot{\mathbf{F}}_{ext} + O(\tau \bar{v} a), \tag{3.15}$$

which, instead of the problematic term involving the derivative of the acceleration, contains a term in the derivative of the external force.

The advantage of this approach is that it avoids some of the troubling aspects of the Lorentz–Dirac equation: The derivation does not involve subtracting out an infinite self-energy; the equation does not predict that charges preaccelerate; and there are no runaway solutions. On Flanagan, Quinn, and Wald's view, however, the regularized equation of motion is only an approximation. Thus, strictly speaking, like the Lorentz force equation of motion, it is inconsistent with the Maxwell equations and energy conservation. The value of that approximation is that it shows that the conceptual problems of an exact classical equation of motion with radiation reaction would surface only in a domain that is well within what, for independent reasons, we take to be the domain of quantum physics.

Unlike other proponents of a regularized equation, Rohrlich argues that the relativistic analogue of (3.15) is an exact equation of motion for point particles, or for what he calls "quasi-point charges" (2001, 2002). For, Rohrlich argues, a strict point charge is impossible in classical electrodynamics. The only legitimate notion of a 'point charge' is that of a charged sphere with a small radius a, whose equation of motion is expanded in powers of a, dropping all terms that depend on positive powers of a. The result is the Lorentz–Dirac (L–D) equation for a "quasi-point charge" (Rohrlich 2002). Rohrlich then points out that *applications* of the L–D equation ought to be governed by the same order of approximations as its *derivation*: "Equations valid only to first order in τ_0 cannot yield solutions valid to higher than first order in τ_0" (Rohrlich 2002, 308). Thus, since any differences between (3.15) and the L–D equation are of higher order, the two equations agree *exactly* within the domain of the L–D equation. Rohrlich concludes: "The problem of the correct dynamics of a classical point charge, first addressed by Lorentz and by Abraham more than a century ago, has finally been given a satisfactory solution" (2002, 310).

Now I believe that Rohrlich is correct in insisting on the importance of validity limits of physical theories. But I do not think that whatever sense it is in which he takes the regularized equation of motion to give the "correct" and "exact" dynamics for charged particles, is sufficient for the purposes of those who insist on the *logical* consistency of a theory's fundamental equations. For while Rohrlich has shown that there is an equation of motion for charged particles that is conceptually less troublesome than the Lorentz–Dirac equation, and is consistent with the Maxwell equations and energy conservation *to whatever degree of accuracy is appropriate in the domain of classical physics*, he has not shown that the theory delineates a coherent class of classical possible worlds. In fact, Rohrlich is faced with the following dilemma: He can assume that the L–D equation is only an approximation (since charged particles ought not to be understood as strict point particles, but only as "quasi-points"). Then his argument for the equivalence of the L–D equation and (3.15), within certain limits, appears to go through, yet from the standpoint of someone who believes that theories have to be internally logically consistent, Rohrlich has given the game away at the very beginning. For the L–D equation is *not* fully consistent with the Maxwell equations and energy conservation for particles that are not strictly pointlike. Or Rohrlich can assume that the L–D equation is an exact equation of motion for point charges. But then the argument for a "strict" agreement between the L–D equation and the regularized equation no longer goes through. For even though the dynamics for a strict point charge is not structure-dependent, if the equation is taken to be exact, it does not follow that all terms of the same size as structure-dependent 'corrections' are zero. That is, if the L–D equation itself is taken to be exact, there no longer is an argument for ignoring the differences between it and the regularized equation of motion (3.15).

Finally, the regularized equation of motion is itself not free of interpretive problems. According to the equation, the instantaneous acceleration of a charge depends on the *derivative* of the instantaneous force in addition to the force itself. This implies in particular that the acceleration at an instant can be nonzero even though the force is zero. This means, contrary to what Rohrlich suggests, that (3.15) violates Newton's first law.[15] A pressing question for defenders of the regularized equation is whether this dependence of the acceleration on the derivative of the force ought to be interpreted causally, departing from the traditional view that only forces act as causes of acceleration. Of course, one might try to avoid this issue by rejecting *any* 'weighty' causal interpretation of the dependence of the motion of charges on electromagnetic fields. But if the backward dependence in the equation expresses a mere functional dependency, it becomes less clear why the Lorentz–Dirac equation is in need of 'regularization' in the first place.

4. Modeling Particle Motions

I want to sum up the discussion so far. I began in the last chapter with a discussion of the standard approach to classical electrodynamics, which treats charged particles as point particles governed by the Newtonian Lorentz force equation of motion. As we have seen, this approach is straightforwardly inconsistent, despite the fact that empirically it is extremely successful and exhibits a number of theoretical virtues: It allows us to model classical electromagnetic phenomena extremely accurately; its fundamental equations are simple; and it fits extremely well into a classical deterministic causal conception of the world. This alone, independent of the subsequent discussion, is, I believe, a point worth making. It shows that, contrary to what often seems to be believed, consistency is not a privileged condition among the various criteria of theory choice, in the sense that consistency is not a necessary condition for a theoretical scheme's being successful, and a scheme can score high on any number of criteria of theory evaluation without being consistent.

But I also claimed something stronger than that—I claimed that the inconsistent point particle scheme is (for all we know) an end product of scientific theorizing which is unlikely ever to be replaced with some fully consistent scheme for modeling classical phenomena involving charged particles. My discussion in this chapter was intended to support this stronger claim. I examined various ways in which one might try to understand the standard theory as an approximation to a more fundamental classical theory. The main question I was interested in asking was Is there a candidate for a conceptually unproblematic and consistent classical theory, from which the standard scheme could be derived as approximation? And the upshot of our discussion is that there seems to be no such theory that can function as a fundamental, microscopic classical electrodynamics and from which the standard theoretical scheme for modeling electromagnetic particle phenomena can be derived.

The Maxwell–Lorentz equations consistently describe the behavior of continuous charge distributions. But a theory of a charged dust does not allow us to model phenomena involving charged particles; and compact localizations of charges cannot consistently be modeled by the Lorentz force equation of motion because of the finite radiation fields associated with accelerating finite charges. One might nevertheless wish to take the position that the continuum theory is the fundamental classical theory and maintain, without further specifying an equation of motion, that charged particles are to be understood as some reasonably well localized maxima in the density of a continuous distribution. These maxima, one might hold, move in accordance with the collective effect of the Lorentz force on its different parts, in conjunction with some further, unspecified constraints which ensure that the charge distribution retains its shape so that the 'particles' do not flow apart. But as it stands, such a 'quietist' attitude leaves us without an equation of motion for charged particles and does not allow us to do much physics. In order actually to represent the behavior of charged particles, the continuum theory has to be augmented and we need to make additional, more substantive assumptions, such as the ones made in the approaches to particle electrodynamics we discussed above.

I examined three approaches to modeling the motion of finitely charged microscopic particles, including radiation reaction effects: Dirac's point particle theory with the Lorentz–Dirac equation of motion; the delayed differential-difference equation for a relativistically rigid charged sphere; and the regularized equation of motion for a localized charge distribution of arbitrary shape. None of these approaches is completely unproblematic, and there is no agreement among physicists as to which of the three is the most promising candidate for a fundamental classical theory of charged particles. The best candidate for an equation of motion that is fully consistent with the Maxwell equations and the principle of energy–momentum conservation arguably is the Lorentz–Dirac equation. Its underlying point particle ontology also recommends itself due to its simplicity, and it fits well with a quantum mechanical understanding of microscopic electric charges, according to which charges do not have any intrinsic structure. But the theory faces the problem of infinite self-fields, and even after renormaliziation the equation of motion is conceptually deeply problematic. Theories of extended charges avoid the infinities, and there are equations that, under the assumption that the electron radius is large enough, avoid the conceptually problematic features of the Lorentz–Dirac equation. Yet these advantages come at a price, for a model of a charged particle fully consistent with a relativistic classical field theory would have to have an infinite number of internal degrees of freedom, and all particle models that have in fact been studied rely on a number of approximations. There is no known model of extended charged particles that is *fully* consistent with special relativity and the Maxwell equations.

In examining the various attempts to derive an equation of motion for charged particles, we see two aspects of physical theorizing at work—two aspects which, I believe, are often in conflict with one another: the *foundational* aim to provide a coherent account of possible ways the world could be, and the *pragmatic* aim to arrive at a practical, useful formalism. Above, I contrasted the role of additional approximation assumptions in electrodynamics with that of approximations in modeling a pendulum. The latter (perhaps overly simplistic) example seems to support a distinction between, on the one hand, a foundational project with the aim of arriving at exact fundamental laws governing phenomena in a certain domain and, on the other hand, the more pragmatic project of applying these laws in modeling specific phenomena. Only the second, according to this distinction, is concerned with practical or pragmatic questions such as that of the mathematical tractability of an equation of motion. While pragmatic considerations can legitimately justify the use of the small angle approximation, such considerations would be misplaced, on this view, as part of the foundational project. I think that some such distinction is implicit in much philosophical theorizing about science. Where one can draw the distinction, no conflict or trade-off between the foundational aim of providing a coherent account of what is physically possible and pragmatic considerations need arise. For the foundational project is logically prior. The fundamental laws, on this view, delineate coherent classes of possible worlds and in principle completely determine the behavior of systems in the theory's domain. Only where the equations governing real systems would be absurdly complicated, do we introduce approximations, idealizations, and abstractions to derive practical, useful representations of individual phenomena.

I do not believe, however, that a sharp distinction between a 'pure' foundational project and one of pragmatic 'model-building' can be drawn in theorizing in classical electrodynamics. Rather, even at the highest theoretical level of deriving an in some sense principled and general equation of motion governing the behavior of charged particles, pragmatic considerations enter. Recall that the derivation of the differential-difference equation proceeds from the assumption that charged particles are relativistically rigid. This assumption appears to be motivated by trying to achieve the best balance between consistency and pragmatic usefulness: The model satisfies the constraints imposed by the special theory of relativity as much as possible, while still being mathematically tractable. Since even the assumption of relativistically rigid particles leads to an impossibly complex equation of motion, there seems to have been little or no interest among physicists in trying to arrive at an equation of motion for fully relativistic charged particles with internal vibrational degrees of freedom.

In choosing a charge distribution for the discrete particles, two simple possibilities suggest themselves: a uniform volume distribution and a uniform surface distribution. Since a volume distribution leads to an infinite series that is too complicated to be summed, a surface distribution is chosen (Rohrlich 1999, 2). But even in this case the resulting infinite series is too complicated to be summed *exactly*, and thus nonlinear terms are dropped. Again, this move reflects a balance of strictness in derivation and pragmatic concerns. The nonlinear terms, one argues, are small enough that they do not affect the motion of the charge considerably. But notice that the differential-difference equation is not obtained by simply dropping all terms smaller than a certain order of magnitude, for one retains linear terms smaller than some nonlinear terms that are dropped. And the reason for retaining *those* terms is simply that this allows us to sum the series and obtain a finite expression for the acceleration of a charge.

Despite the approximations involved in its derivation, physicists who endorse the differential-difference equation of motion take it to be the *basic* or fundamental equation governing classical extended charged particles (see, e.g., Rohrlich 2000, 12). The equation is taken to be basic, despite the fact that it is, strictly speaking, inconsistent with the Maxwell equations, energy conservation, and the special theory of relativity. And even though the infinite series from which the differential-difference equation is derived contains fewer approximations than the latter equation, it is the latter and not the former that is generally regarded as *the* equation of motion for extended charged particles. An obvious explanation for this is that simplicity is an important criterion in theory assessment. The infinite series is far too complex to be acceptable as a fundamental equation of motion. By contrast, even though the differential-difference equation takes us farther from the Maxwell equations, and thus in some intuitive sense might be thought to be less consistent with these equations than the infinite series, this equation achieves a better balance of simplicity, consistency, and accuracy. Thus, the treatment of the delayed differential-difference equation supports my contention that logical consistency is not a privileged criterion of theory assessment, but is only one of a range of criteria among which physicists aim to achieve a balance.

The derivation of the regularized, structure-independent equation of motion is similarly rife with approximative assumptions. The main motivation for this approach is pragmatic: Since there is no empirical evidence favoring certain assumptions concerning the structure of the electron over any others, we should, according to this approach, adopt an equation that abstracts from any such specific assumptions. And since both the Lorentz–Dirac equation and the regularized equation of motion are (from the perspective of an extended particle theory) valid to the same degree of approximation, one is free to choose that equation which seems conceptually less problematic. And the same point as in the case in the previous paragraph applies here: It is the simple regularized equation of motion, which is derived with the help of approximations, rather than the complex infinite series expression, which is regarded as the correct equation of motion (see Rohrlich 2002).

As philosophers we might be tempted to think that physicists are simply confused when they speak of an approximate equation as "fundamental," "correct," or even "exact." This, however, would mean imposing a philosopher's conception of theories on science rather than trying to understand the practice of theorizing. Instead, I want to propose that we should read such claims as indications for which sets of equations physicists themselves take to be the most basic and important in a certain domain, and then ask what criteria of theory choice would allow us to make sense of the physicists' decisions. Clearly, when Rohrlich speaks of the differential-difference equation or of the regularized equation of motion as "correct" or "exact," he cannot be committed to the claim that these equations are *logically* consistent with the Maxwell equations and the principle of energy conservation from which they are derived. But instead of claiming that this shows Rohrlich is confused, I want to propose that we adopt a principle of charity and interpret the physicists' claims in a way that renders them defensible. And on any such interpretation, I take it, internal consistency does not come out as a necessary condition governing theory choice, since considerations of simplicity, mathematical tractability, and broad conceptual fit appear to be able override concerns for strict logical consistency.

The only classical equation of motion that has a claim to being strictly consistent with the Maxwell equations and energy–momentum conservation is the Lorentz–Dirac equation with all its conceptual problems. However, those who endorse it, as Rohrlich did at one time, insist that it is useful only if one does not look too closely, as it were. Up to a certain degree of approximation, the backward causal and nonlocal behavior predicted by the equation does not show up, and the theory is compatible with a standard causal conception of the world. Interestingly, approximation proceeds in the opposite direction in arguments for the Lorentz–Dirac equation than one might expect. Commonly approximation is understood as taking us away from the 'truth': While we think of Newton's equation of motion as strictly governing pendulums, we take the small angle approximation to lead to a less accurate (even if more useful) representation. Of course we do not believe that Newton's laws are true laws—a true theory would presumably be some quantum-gravitational theory—but as far as classical phenomena are concerned, Newton's laws are in some sense the correct laws. In particular, no 'coarse-graining' of the laws will get us closer to a classically correct description; rather, any approximation to the theory will, within its domain, get us farther from the truth—or at least so the story goes. By contrast, Rohrlich's justification of the Lorentz–Dirac equation presupposes that classical electromagnetic systems are in fact forward-causal and causally local; he then argues that the backward-causal equation is nevertheless acceptable, because a suitable approximation to the equation does not conflict with this presupposition. Thus, the assumption is that coarse-graining or approximating the fundamental equation of motion takes us closer to the truth.

Reasoning about approximate fit plays a role in the justification of the Lorentz–Dirac equation in ways that one might more commonly associate with theory applications. Those who accept the equation as the fundamental equation of motion of classical electrodynamics see it as striking a good balance between rigor in its derivation from the Maxwell equations and fit with overall conceptual constraints. Those who do not accept it weigh the criteria of theory assessment differently. For example, the fact that the regularized equation of motion fits better into our preferred causal framework might be taken to outweigh the fact that the equation is not fully consistent with the Maxwell equations. In all this, physicists, it seems, are relatively unconcerned with the philosopher's questions of whether the theory delineates a coherent class of physically possible worlds. Of course, all other things being equal, a consistent fundamental theory is preferable to one that is not. But when full consistency is not attainable, this does not mean defeat. Rather, the focus seems to be on arriving at a set of basic yet practically useful equations that within the theory's domain of applicability accurately represent the phenomena in its domain.

It seems to me that one could take our investigation of classical electrodynamics to suggest that there can be a disparity or mismatch even between our best mathematical models and the physical phenomena we want to represent. Many classical electromagnetic phenomena are best represented as involving an interaction between discrete particles and continuous fields. Yet, as we have seen, there does not seem to be a completely satisfactory and consistent mathematical formalism through which one can model such interactions. Often the point particle ontology with its resulting infinities seems to be blamed for the conceptual problems of the theory. But it seems to me that the infinities alone cannot be what is at fault, since the point particle equation of motion is problematic even after it is renormalized. The problem seems to lie at least as much with the fact that fields and charges mutually determine each other; it is this 'feedback' effect that seems to render the theory deeply problematic. Some of the problems in representing interactions between localized, discrete objects and continuous fields seem to re-emerge in quantum field theories. The best tools for representing many electromagnetic phenomena are provided by particle–field theories, yet these theories do not provide us with satisfactory 'global,' unified, and coherent representations of all the phenomena in their purview.

5. Conclusion

In this chapter I investigated equations of motion that are alternatives to the standard Maxwell–Lorentz theory. I argued that there is no consistent and conceptually unproblematic theory that covers particle–field phenomena. The only consistent and relatively 'well-behaved' electromagnetic theory is not a particle theory at all—the theory of charged dusts. Yet this theory on its own is incompatible with the existence of charged particles. I also argued that the kind of equations that physicists propose as 'fundamental' or 'exact' equations, and the reasons physicists offer in support of these equations, suggest that there is no sharp distinction between a foundational project aimed at finding coherent and possibly true representations of what is physically possible and a project concerned with practical, useful representations of particular phenomena. For even arguments for putatively fundamental equations frequently involve approximation procedures motivated by appeals to pragmatic concerns such as simplicity.

4

Non-Locality

1. Introduction

Classical electrodynamics is generally understood to be the paradigm of a local and causal physical theory. In fact, one of the main motivations for interpreting electromagnetic fields realistically appears to be that a theory with real fields promises to be a local theory. If fields transmit the actions of electric charges on one another, then electrodynamics ought to be able to avoid some of the conceptually unsatisfactory features of Newtonian gravitational theory. In Newton's theory, massive bodies exert forces on one another instantaneously, across gaps in space. Positing real fields that transmit the influence of one charge on another at a finite speed promises to avoid such action-at-a-distance.

But what, exactly, is it for a theory to be local or nonlocal? This is the question I want to ask in this chapter. As I will argue, there are several logically distinct locality conditions in classical physics which unfortunately are often not distinguished carefully enough. Informally, locality principles are often introduced in causal terms. Newtonian gravitational theory, for example, is said to be nonlocal, because it allows for action-at-a-distance, while the theory of special relativity is often said to imply the locality condition that there can be no superluminal causal propagation. According to a widespread view, however, the notion of causation should have no place in fundamental physics, and whatever genuine content such prima facie causal principles have, should be explicated in noncausal terms. Thus, Bertrand Russell famously argued that in the advanced sciences the notion of functional dependency has replaced that of causation. Any more substantive notion of causality, he claimed, "is a relic of a bygone age, surviving, like the monarchy, only because it is erroneously supposed to do no harm" (Russell 1918, 180). I want to argue here that Russell was wrong. While the dynamical laws of classical electrodynamics do not imply any 'rich' causal relations, the notion of cause does play an important role in *the interpretive framework* of this and other theories. And whatever notion is employed there cannot be reduced to that of functional dependency. For, I will argue, some of the locality conditions invoked in characterizing classical theories are irreducibly causal.

As my main case study I will appeal to Dirac's classical theory of the electron (Dirac 1938), which we have already encountered in chapter 3. In light of the important role that principles of locality have played in the interpretation of classical electrodynamics, it might come as a surprise that Dirac's theory, which, as we have seen, is the most promising candidate for a fully consistent theory of classical charged particles, is causally nonlocal. On its standard interpretation, the theory allows for forces to act where they are not and for superluminal causal propagation.[1]

In the next section I will discuss two Russellian worries one might have about the notion of cause. I will distinguish two prima facie causal locality conditions and a condition that might be thought to offer a noncausal explication of one or the other of the two causal conditions. In section 3 I will examine how appeals to locality conditions affect the interpretation of classical electrodynamics, with respect to the roles of electromagnetic fields and potentials in the theory. Then I will turn to Dirac's theory of the electron and argue that the theory is best interpreted as being causally nonlocal in two senses. As part of the argument we will have to examine the asymptotic condition of vanishing acceleration at infinity, which plays an important role in the theory, in some detail. In sections 5 and 6, I will appeal to examples of 'respectable' scientific theories (including Dirac's theory) to argue that the three main locality conditions I have identified are all independent from one another. In section 7, I discuss one additional candidate for a noncausal explication of the causal conditions—a condition proposed by Earman (1987)—that is meant to capture the content of the action-by-contact principle. Section 8 discusses a possible objection to my claim that causal locality principles play an important role in fundamental physics.

The claim I will be arguing against is not that the various locality principles logically entail one another. Simple inspection can show that they do not. Rather, I am interested in the weaker claim that there is some close conceptual connection between the various principles such that any theory seriously entertained by scientists which satisfies one of the locality criteria satisfies all others, and that any such theory that violates one criterion violates all others.

2. Causes and Locality Conditions

Why might one think that the notion of cause is not needed in fundamental physics, or might even be incompatible with it? One of Russell's worries appears to have been that the fundamental ontology of physical theories appears to consist of point events, such as that of a point charge having a certain position and velocity at some time t or that of the electromagnetic field strength's having a certain magnitude at a certain space-time point. But an integral feature of the notion of causation, according to Russell, is that causes have direct effects to which they are contiguous. With David Hume, Russell assumed that part of the notion of cause is that causes satisfy a contiguity condition: Causes are 'next to' their immediate effects. However, given the density of the real numbers, there is an infinity of points between any two points, and hence no two points are contiguous to one another. If the fundamental events of physics are pointlike, then these events cannot satisfy the causal contiguity condition that causes are spatially contiguous to their direct effects.

One might reply to Russell's worry that it is far from obvious that the contiguity condition really is part of the concept of cause. There are conceptions of causation, it seems, such as the intuitive notion of a cause 'bringing about' its effects, that appear to be compatible with causes acting at a distance. The concept of 'gappy' causation does not seem to contain a contradiction. Yet even if the notion of causation-at-a-distance is not contradictory, such a notion has struck many as problematic. Russell's worry shows that unless we want to accept that *all* causation is at a distance, we have to find a way to spell out what it is for causes to act locally that does not require the contiguity condition to hold. But it seems that this challenge is easily met. Instead of requiring causes to have direct effects to which they are contiguous, we can stipulate that locally acting causes are those which are connected to their effects through a continuous causal process that mediates any interaction between a cause and any of its spatially distant effects.

This idea appears to be captured, for example, in Albert Einstein's principle of *local action*, which he introduces in the following passage:

> Characteristic for... physical things is that they are thought of as being arranged in a spatiotemporal continuum. Furthermore it appears to be essential for this arrangement of the things introduced in physics that these things claim an existence independent of one another at a specific time, insofar as these things "lie in different parts of space." Without the assumption of such an independence of existence (the "being such-and-such") of spatially distant things, which originally arose out of everyday thinking, physical thinking, in the sense familiar to us, would not be possible. Without such a clear separation one also cannot grasp how physical laws can be formulated and tested. Field theory has carried this principle to the extreme, in that it localizes within infinitely small (four-dimensional) space-elements the elementary things existing independently of one another that it takes as basic, as well as the elementary laws it postulates for them.
>
> For the relative independence of spatially distant things (A and B) this idea is characteristic: externally influencing A has no *immediate* influence on B; this is known as the "principle of local action," which is applied consistently only in field theory. Abandoning this principle completely would render impossible the idea of the existence of (quasi-) closed systems and thereby make it impossible to postulate empirically testable laws in the sense familiar to us. (Einstein 1948, 321–322)[2]

As Richard Healey (and others) have pointed out, Einstein appears to appeal to two distinct locality principles in this passage. One is the principle of *local action*: "If A and B are spatially distant things, then an external influence on A has no immediate effect on B" (Healey 1997, 23). The other is a principle Healey calls *separability* and expresses as follows: "Any physical process occurring in spacetime region R is supervenient upon an assignment of qualitative intrinsic physical properties at spacetime points in R" (Healey 1997, 27). Intuitively, intrinsic properties are properties an object has in and of itself, without regard of other things; and "a process consists of a suitably continuous set of stages, typically involving one or more enduring systems" (ibid., 28). The separability principle is meant to capture the idea that the state of any system is completely determined by the states of its localized subsystems.

The principle of local action demands that any effect of one pointlike event on another be mediated by a continuous causal process. At least this is one possible reading of the principle. As Healey notes, the principle is ambiguous in that the term 'immediate' might be read as either 'unmediated' or 'instantaneous' (in the case of nonrelativistic theories).[3] Healey maintains, however, that this ambiguity "seems relatively harmless, in so far as any instantaneous effect would have to be unmediated" (Healey 1997, 24). This is a mistake. As we will see below, there are examples of perfectly respectable scientific theories which postulate that effects are felt instantaneously—that is, that causal processes propagate infinitely fast—and at the same time stipulate that all causal influences have to be mediated. Thus, we ought to distinguish between two different locality principles—the first stating that all causal interactions are mediated by continuous causal processes or continuous 'chains' of causes, and the second demanding that all causal influences propagate at a finite speed.

Analogous to the condition that all causal influences between *spatially* separated things are mediated, we can demand that *temporally* separated things or events that interact causally also be linked by an unbroken causal chain. Thus, in a proposal for making precise the prohibition against unmediated causation between events that do not have the same spatiotemporal location, Lange distinguishes three conditions—spatial and temporal locality, as well as spatiotemporal locality. For our purposes it will sufficient to focus on the last condition:

> *Spatiotemporal locality:* For any event E, any finite temporal interval $\tau > 0$ and for any finite distance $\delta > 0$, there is a complete set of causes of E such that for each event C in this set, there is a location at which it occurs that is separated by a distance no greater than δ from a location at which E occurs, *and* there is a moment at which C occurs *at the former location* that is separated by an interval no greater than τ from a moment at which E occurs *at the latter location*. (Lange 2002, 15; italics in original)

Spatiotemporal locality (which is logically stronger than the conjunction of spatial locality and temporal locality) requires that for any event E, there is at each moment, separated from E by a time τ, a complete set of causes within a finite region of radius δ surrounding the location of E's occurrence, and this holds even as $\delta \to 0$ and $\tau \to 0$.[4] Since Lange's principle has the two advantages of avoiding the ambiguity in Einstein's principle and of covering the case of temporally distant things as well, I will appeal to Lange's principle in what follows.

I suggested that we should distinguish the condition that all causal interactions propagate at a finite speed from the condition of spatiotemporal locality. In fact, there are several different locality conditions concerning the velocity with which effects propagate. Earman (1987, 451) distinguishes the principle that all causal propagation takes place with a finite velocity from the strictly stronger principle that there is a fixed, finite limiting velocity for all causal propagation. The condition that there is no superluminal causal propagation is a specific version of the latter principle. We can call this last principle *relativistic locality*.[5]

Relativistic locality prohibits an event's being a cause of any other spacelike separated event. Moreover, for each principle concerned with the speed of causal propagation there is a corresponding principle about *signaling* speeds, such as the principle that there is no superluminal signaling. Since signaling is a type of causal process, but there may be causal processes that cannot be exploited to send signals, the principles concerning causal processes are strictly stronger than the respective signaling principles. The differences between these conditions are of no importance to what follows, and for ease of exposition I will continue to speak as if there were just one locality condition restricting the velocity with which causal propagation takes place.

A second (perhaps also Russellian) objection to employing the notion of cause within the context of fundamental physics, points to the fact that the causal relation is asymmetric. Often this asymmetry is characterized in temporal terms and captured in the demand that causes ought not to follow their effects. But even in cases that suggest a backward-causal interpretation, such as that of the Lorentz–Dirac equation, the relation is asymmetric: Nonzero field forces on the worldline of a charge in the future are taken to be a cause of the charge's acceleration now, but not vice versa. However, any such asymmetry, the objection argues, is not underwritten by the basic equations of our fundamental physical theories. The basic equations of our most fundamental classical theories are time-symmetric differential equations and are deterministic in both directions. That is, the laws, together with a complete specification of the state of the world at one time, determine the state of the world at all other later, as well as *earlier*, times.

Now, this symmetry of determinism in the laws of classical physics does not appear to be *incompatible* with adding an asymmetric causal structure to the laws. But on one influential view of the nature of scientific theories, adding a causal superstructure that is not implied by a theory's fundamental equations is illegitimate. This view, which we discussed in chapter 1, holds that physical theories consist of a mathematical formalism (or a class of mathematical models) and an interpretation whose *only* job it is to fix the ontology of the theory. Absent from the account is the idea that part of the job of an interpretation may be to stipulate causal structures consistent with the mathematical models that the theory provides. Causal structures simply do not seem to be needed in fundamental physics. On this view, then, we can make sense of theorizing in the physical sciences without invoking a 'weighty' notion of cause, and neo-Humean scruples advise against including anything in the proper content of a scientific theory that is not implied by the theory's formalism.

What does this mean for the notion of locality? Both Einstein's principle of *local action* and Lange's condition of *spatiotemporal locality* are framed in causal terms. Lange's condition explicitly refers to complete sets of causes of an event E, while Einstein's principle invokes the idea of an *external influence* on a system and that of the influence of one event on another, which are part of a cluster of prima facie causal notions that includes the concepts of intervention and manipulation. Clearly, to say that one thing can have an influence on another is to say more than that there is a correlation between the two things. To cite a familiar example, there can be a correlation between the reading of a barometer and the advent of a storm, yet it would be wrong to say that barometer readings can have an influence on the weather. Tracking the results of (hypothetical) external influences on (or interventions into) an otherwise closed system appears to provide us with a guide to causal relations in the system. In fact, Einstein seems to suggest some such connection when he maintains that the principle of local action is necessary for the idea of a (quasi-)closed system and for the possibility of testing a theory empirically. I will appeal to the connection between the notion of causation and that of counterfactual interventions in chapter 7, where I defend a causal reading of the *retardation condition*, according to which accelerated charges produce diverging disturbances in the electromagnetic field.

Einstein's principle of local action and Lange's principle of spatiotemporal locality show that there are ways to distinguish locally acting causes from causes that act a distance without having to appeal to a notion of a cause being 'next to' or 'immediately preceding' its direct effects. Thus, Russell's first objection can be met. Moreover, the principles seem to draw meaningful and scientifically useful distinctions. Thus, if we were to accept the second objection against causes and wanted to rid physics of any 'weighty' concept of causation, we would have to try to explicate the principles in purely noncausal terms, presumably as principles of determinism.

One such principle was proposed by Erwin Schrödinger as a noncausal explication of the prohibition against unmediated action-at-a-distance. Since Schrödinger, like Russell, apparently took this prohibition to be at the core of the notion of causation, he called his principle the "principle of causality": "The exact situation at *any* point P at a given moment is unambiguously determined by the exact physical situation within a certain surrounding of P at any previous time, say $t - \tau$" (Schrödinger 1951, 28). Schrödinger adds that "the 'domain of influence' [that is, 'the surrounding' of P] becomes smaller and smaller as τ becomes smaller." "Classical physics," he claims, "rested entirely on this principle" (ibid., 29). Despite its name, Schrödinger's principle is, of course, not a causal principle in any 'weighty' sense, and ought to be acceptable to Russellians or neo-Humeans. A similar, if somewhat broader, notion of 'cause' is also endorsed by Niels Bohr: "In physics, causal description ... rests on the assumption that the knowledge of the state of a material subsystem at a given time permits the prediction of its state at any subsequent time" (Bohr 1948, 312).

According to Schrödinger and Bohr, the notion of cause can be reduced to a notion of determinism. Both characterize causation asymmetrically, yet the claim that causes precede their effects is, as in Hume's theory, purely a matter of definition and does not reflect any physical asymmetry. For Bohr, a theory is causal simply if it is forward-deterministic, while Schrödinger's principle adds the constraint that the state of a system is locally determined by its past. Yet missing in such a neo-Humean conception of causality is any tight connection between the notion of causation and those of manipulability and control that may seem to be implicit in Einstein's principle of local action. Thus, we could define a principle of 'retro-causality' in analogy to Schrödinger's principle, and the theories with time-symmetric laws with which we are familiar would obey both Schrödinger's and the retro-causal principle. And nothing of any real significance seems to hang on our reserving the name 'causation' for the forward-looking principle.

More recently, Gordon Belot distinguished two noncausal locality conditions:

(i) *Synchronic Locality*: the state of the system at a given time can be specified by specifying the states of the subsystems located in each region of space (which may be taken to be arbitrarily small).
(ii) *Diachronic Locality*: in order to predict what will happen *here* in a finite amount of time, Δt, we need only look at the present state of the world in [a] finite neighbourhood of *here*, and the size of this neighbourhood shrinks to zero as $\Delta t \to 0$. (Belot 1998, 540)

The first condition is a version of the separability principle and is meant to capture the nonholist intuition that the properties of a (classical) system ought to be reducible to the properties of its parts. The second condition is equivalent to Schrödinger's principle. Like Schrödinger's principle, Belot's principle is time-asymmetric, but the asymmetry does not have any physical significance. Unlike Schrödinger, however, Belot suggests that diachronic locality is equivalent to the condition that causal influences travel at a finite speed: Newtonian gravitational theory is diachronically nonlocal, Belot says, "since gravitational effects propagate with infinite velocity" (Belot 1998, 541), while classical electrodynamics is diachronically local, since "electromagnetic radiation propagates at a fixed speed" (ibid.). Belot also claims that diachronic locality implies synchronic locality.[6] Since he says, further, that if a magnetic field were allowed to act where it is not, synchronic locality would be violated, it follows that he is committed to the claim that the condition of diachronic locality implies the condition that all causes act locally.

In what follows, I will investigate in some detail the relations between the condition proposed by Schrödinger and Belot, on the one hand (for which I will adopt Belot's label "diachronic locality"), and the two explicitly causal locality conditions of a finite speed of causal propagation and of spatiotemporal locality, on the other. In the next section I will discuss some ways in which appeals to locality conditions influence the interpretation of classical electrodynamics.

3. Locality, Fields, and Potentials

Classical electrodynamics describes the interaction between charged particles as being mediated by electromagnetic fields. One of the main motivations for interpreting the electromagnetic field realistically appears to be the aim of arriving at a theory that is local in some sense. And in fact, the microscopic Maxwell–Lorentz theory familiar from physics textbooks, which I discussed in chapter 2, is local in every one of the senses I distinguished in the last section: The theory is diachronically local; disturbances in the electromagnetic field propagate at a finite speed; and, according to the standard causal interpretation, fields constitute spatiotemporally local causes for the acceleration of charged particles. It is this theory, then, that people seem to have in mind when they refer to electromagnetism, as Belot does, as "the paradigm of all that a classical (i.e. non-quantum) theory should be" (1998, 531).

Another important motivation for interpreting fields realistically is that a theory in which fields are understood as mere calculational devices violates energy and momentum conservation (see Lange 2002). Since electromagnetic interactions between charged particles take a finite time to propagate, the action of one charge on another need not be balanced by an equal and opposite reaction by the other charge, and the energy of the particles alone will not be conserved. The hope is that energy and momentum conservation can be satisfied if the energy and momentum change in a charge is balanced locally by a change in the field's energy and momentum. Now, one problem with both these lines of argument for the reality of fields is that they raise the question as to what our grounds are for believing in certain locality conditions, on the one hand, and energy conservation, on the other.[7] A second problem concerns the question of whether classical electrodynamics, even with real fields, does in fact result in a local theory in which energy and momentum are conserved. We have already seen that the status of the principle of energy conservation is deeply problematic in the theory. According to the Maxwell equations, accelerated charges radiate energy, but the Lorentz law governing the motion of charged particles ignores any effects on the motion of a charge due to its own radiation. Thus, the Maxwell–Lorentz equations are inconsistent with the principle of energy conservation, despite the fact that the principle is an important component of the theory.

In this chapter we will see that the most promising candidate for a consistent classical particle–field theory, Dirac's theory, is nonlocal in the two causal senses distinguished above. Thus, Lange's conclusion that, given that the electromagnetic field can be interpreted realistically, "spatiotemporal locality is thus (at last!) secured for electromagnetic interactions" (Lange 2002, 247) needs to be qualified: Electromagnetic interactions are causally local *only* if they are modeled within the standard, inconsistent Maxwell–Lorentz theory, but not in Dirac's theory, which consistently tries to include radiation reaction effects in the particle equation of motion.

In addition to the electromagnetic field strengths, the mathematical formalism of classical electrodynamics makes use of the *vector* and *scalar potentials*, which are related to the fields via

$$\mathbf{B} = \nabla \times \mathbf{A} \tag{2.3}$$

and

$$\mathbf{E} = -\nabla \Phi - \frac{1}{c}\frac{\partial \mathbf{A}}{\partial t}. \tag{2.4}$$

The potentials can be combined to a four-potential $A^\alpha = (\Phi, \mathbf{A})$, in terms of which the field tensor (with the electric and magnetic field strengths as its components) becomes $F^{\mu\nu} = \partial^\mu A^\nu - \partial^\nu A^\mu$.

In the standard interpretation of the theory, the potentials are treated as mere mathematical fictions. They are introduced for reasons of mathematical convenience, but are not taken to have any physical significance beyond the fields that are determined by them. Two reasons are offered for not interpreting the potentials realistically. The first is that the potentials are not determined uniquely by the observable fields $\mathbf{E}$ and $\mathbf{B}$, but only up to a *gauge transformation*, $\mathbf{A} \rightarrow \mathbf{A}' = \mathbf{A} + \nabla \Lambda$ and $\Phi \rightarrow \Phi' = \Phi - (1/c)(\partial \Lambda/\partial t)$, where Λ is an arbitrary scalar function. That is, two sets of potentials $\mathbf{A}, \Phi$ and $\mathbf{A}', \Phi'$ that are related by a gauge transformation determine the fields **E** and **B**. Thus, the potentials are what Lange calls "a dangler" in the theory (Lange 2002, 47). Their absolute values cannot be measured, and make no difference to any other physical fact. All that does make a difference is the gradient of the scalar potential and the curl and time derivative of the vector potential—that is, roughly, how the potentials change and differences in the potentials. If we do not want our theories to include such danglers, which (at least by the theory's lights) are physically irrelevant to anything else that occurs, we should not interpret potentials realistically.

The second reason offered is that if we interpret the potentials realistically, the theory will no longer be deterministic (see Belot 1998; Healey 2001). The Maxwell equations can be rewritten as a partial differential equation for the evolution of the potentials. The definitions of the potentials imply that the two homogeneous Maxwell equations are satisfied automatically. The inhomogeneous equations in four-vector notation are

$$\partial_\mu(\partial^\mu A^\nu - \partial^\nu A^\mu) = j^\nu. \tag{4.1}$$

This equation, according to the argument, does not determine how the potentials evolve from given initial data in the potentials, but only up to a gauge transformation. For there are infinitely many gauge transformations that are the identity on any given initial-value surface but diverge smoothly from the identity elsewhere. Thus, determinism fails: The state of the world at some time, together with the laws of nature, does not uniquely determine the state of the world at all other times.

One should note, however, that the argument cited by Belot and Healey presupposes that what the laws of the theory are, is fixed independently of and prior to the ontology of the theory.[8] If we have fixed the laws to be *only* the Maxwell equations, then interpreting potentials realistically results in an indeterministic theory. Yet we might also have fixed the theory's ontology first and then asked whether there are laws which result in a deterministic theory. And the answer to this question appears to be that there are such laws. For example, a theory that in addition to the Maxwell equations had the Lorenz gauge condition $\partial_\mu A^\mu = 0$ as a law would be deterministic.

Notice that the two arguments against interpreting potentials realistically are independent of each other. The first argument maintains that the absolute values of potentials are not measurable and are not correlated with any other known physical quantity. There is no evidence we could have for determining what the 'true gauge' was, if it existed. The second argument claims that the absolute value of the potential at some time, whatever it may be, together with the values of the relevant derivatives of the potential at that time, does not uniquely determine how the potential evolves in time. This latter objection to real potentials could be met by positing an additional law that fixes the gauge. But this would not help with the first worry; this additional law (and the 'true gauge') would still be unknowable to us.

If we want to exclude danglers from our theories, then we cannot take the electromagnetic potentials to be physically real. Thus, the state of the electromagnetic field is determined completely by the local values of the field strengths **E** and **B** at each space-time point. Belot (1998) argues that considerations from quantum mechanics might put pressure on this interpretation. For the

Aharanov–Bohm effect apparently shows that differences in the vector potentials can have empirically observable consequences even in regions of space where there are no differences in the electromagnetic field. According to quantum mechanics, the interference pattern produced by a beam of electrons passing through a double-slit apparatus may by affected by changes in a magnetic field that is confined to a region from which the electrons are excluded. While the magnetic field can be set to zero everywhere along the path of the electrons, the value of the potential varies along the path in the Aharonov–Bohm effect with the variation in interference patterns. Thus Belot suggests that for reasons of interpretive continuity, we should abandon the standard interpretation of *classical* electrodynamics. If we have to interpret potentials realistically in quantum mechanics, then we ought to do the same in classical physics. The Aharanov–Bohm effect has taught us, he says, that we had "misunderstood what electromagnetism was telling us about the world" (Belot 1998, 532).

In fact, as both Healey and Belot point out, there are two different interpretations suggested by the Aharanov–Bohm effect. We can take the local values of the electromagnetic potential to represent a real physical field. Or, since the interference pattern of the electrons does not depend on the value of the potential at a single point, but on the integral of the potential along a closed curve, we can take the state of an electromagnetic system to be described by the *Dirac phase factor*, which involves integrals of the potential over all closed curves in space-time:

$$\exp\left[-(ie/\hbar)\oint A^\mu(x^\mu)\cdot dx^\mu\right].^9 \qquad (4.2)$$

The advantage of the second interpretation is that the Dirac phase factor is gauge-invariant.[10] Even in quantum mechanics the local values of the potentials are not empirically measurable; only the values of integrals around closed curves are. Thus, the local values of the potential are danglers in quantum mechanics as well. By contrast, if we interpret the Dirac phase factor (but not the vector potential) realistically, we do not need to take 'danglers' on board, and we do not have to give up determinism. But taking the state of the field to be completely specified by the Dirac phase factors has the disadvantage of rendering classical electrodynamics radically nonlocal. For then the intrinsic properties of an electromagnetic system do not supervene on properties of the system at individual space-time points because there are no local physically real quantities out of which the line integrals are constructed. That is, the theory postulates nonseparable processes and violates Belot's condition of synchronic locality.

Belot also claims that interpreting the potentials realistically results in a diachronically nonlocal theory. While this claim is correct, Belot's argument for it is not cogent. As an example of how the potential supposedly depends nonlocally on a change in current density, Belot cites the case of a very long solenoid—a conducting wire coiled around a cylinder. If a constant current is running through the wire, then (according to the Biot–Savart law) there will be a constant magnetic field inside the device, while the magnetic field outside will be zero. Since the holonomy around a closed curve which loops around the solenoid is equal to the magnetic flux through the area enclosed by the curve, Belot argues that "it follows that the vector potential propagates with infinite velocity": "If we switch the thing on, then the values of the vector potential at some point arbitrarily far away must change instantaneously" (1998, 549, fn.). But the Biot–Savart law is a law of magnetostatics and does not apply to time-dependent currents. The math in this case is a lot more complicated, since the magnetic field depends not only on the current density but also on the derivative of the current density, but what we would find is that if we switch the solenoid on at $t=0$, there will be a nonzero electromagnetic pulse spreading out from the solenoid, which ensures that the flux through any area with radius greater than ct is zero.[11]

That interpreting the potentials realistically results in a diachronically nonlocal theory does follow, however, directly from Healey's and Belot's indeterminism argument. Since the values of the potential and its derivatives on a given initial value surface do not determine the value of the potential in the future, we cannot find out by looking here, what will happen to the potentials next (even if the values of the potentials were measurable).

Belot's mistake here might be partly responsible for the fact that he does not see his claim that synchronic nonlocality implies diachronic nonlocality to be in need of much of a discussion. However, on one natural construal of what it is "to predict what will happen *here*," one can predict only how local quantities change *here*, since nonlocal quantities, like the Dirac phase factor, do not "happen *here*" at all. But then synchronic nonlocality does not entail diachronic nonlocality. Since a synchronically nonlocal theory can be concerned with local quantities as well, such a theory can be diachronically local in the sense that what will happen *here* to the values of localized quantities can be determined locally. Moreover, even if we are willing to accept that Dirac phase factors are the kind of things that can happen *here*, electrodynamics is still local across time in an intuitive sense that Belot's criterion does not capture: Even though the phase factors are "spread out" through space, changes in their values do not propagate instantaneously. Since the vector potential propagates at a finite speed, the phase factor around paths far away from a disturbance in the field will change only after a finite time, when the disturbance has reached at least some point on the path.

4. Dirac's Classical Theory of the Electron

4.1. Forces Acting Where They Are Not

In the previous two chapters we discussed several approaches to modeling the interactions between microscopic electric charged particles and electromagnetic fields. In the theory familiar from physics textbooks, electromagnetic phenomena are treated as being governed by two sets of laws—the microscopic Maxwell equations, on the one hand, and the Lorentz force law as a Newtonian equation of motion, on the other. The Maxwell–Lorentz approach to electrodynamics results in a local and forward-causal theory, but it is inconsistent, since the Lorentz law ignores any effects on the motion of a charge due to its own radiation.

As we have seen, however, it is possible to include a charge's radiation field in an equation of motion for a charged particle in an arguably consistent way.

84 Inconsistency, Asymmetry, and Non-Locality

The stumbling block in trying to derive an equation of motion for a point charge from the Maxwell equations and the principle of energy–momentum conservation is the infinity in the charge's field, which implies that the energy associated with the field is infinite. This difficulty can be overcome if we 'renormalize' the mass of the charged particle—that is, if we absorb part of the infinite self-energy of a charge into its mass (which nevertheless is taken to be finite). The result is the Lorentz–Dirac equation, which first was derived by Dirac (1938) and whose nonrelativistic approximation is (3.9) above:

$$ma^\mu = F^\mu + \frac{2e^2}{3c^3}\left(\frac{da^\mu}{d\tau} - \frac{1}{c^2}a^\lambda a_\lambda v^\mu\right).^{12} \tag{4.3}$$

Here F^μ is the total external force (due to the external electromagnetic field $F^{\mu\nu}$ and any nonelectromagnetic forces acting on the charge), the mass term m on the left represents the renormalized finite mass, and derivatives are with respect to the proper time τ. Recall that this equation differs from familiar Newtonian equations in that it is a third-order differential equation, involving not only the acceleration a^μ of the charge but also its derivative.[13] In renormalizing the mass one needs to assume as an asymptotic condition that the acceleration of the charge tends to zero at both future and past infinity. Thus, even though (4.3) is a *local equation* in that it relates quantities at a single proper time τ to one another, it is derived with the help of a *global assumption*.

If we explicitly impose the requirement that all acceptable solutions to the Lorentz–Dirac equation have to satisfy the asymptotic condition of vanishing acceleration at infinity, equation (4.3) can be integrated once and we arrive at the following second-order integro-differential equation of motion (of which (3.13) is the nonrelativistic approximation):

$$a^\mu(\tau) = \int_\tau^\infty e^{(\tau-\tau')/\tau_0}\left[\frac{1}{m\tau_0}F^\mu(\tau') - \frac{1}{c^2}a^\lambda(\tau')a_\lambda(\tau')v^\mu(\tau')\right]d\tau', \tag{4.4}$$

where the constant $\tau_0 = 2e^2/3mc^3$. This is a nonlocal equation in that it relates the acceleration at τ to the acceleration at all other times after τ.

Unless one assumes the asymptotic condition, the mass of the electron cannot be renormalized, since the infinite integrals do not formally look like mass terms. Yet its mathematical indispensability is not the only motivation for adopting the asymptotic condition. Physically the condition can be motivated, first, by the assumption that any interaction between a charge and external fields can be modeled as a scattering process—that is, as an interaction between a charge that is "asymptotically free" and localized external fields—and, second, by appealing to a principle of inertia according to which the acceleration goes to zero sufficiently far away from any force acting on the charge.[14] Even though this principle is weaker than the familiar Newtonian principle, according to which the acceleration of an object is zero if no external force is acting on it, adopting it has the advantage of ensuring a certain interpretive continuity between Dirac's theory and Maxwell–Lorentz electrodynamics (and Newtonian theories in general).[15] For example, in both theories a charge which never experiences a force moves with constant velocity (see Rohrlich 1990, sec. 6.10).[16]

The "obvious" interpretation (Rohrlich 1990, 149) of (4.4) is that in Dirac's theory, as in Newtonian theories, forces should be taken to be causally responsible for the acceleration of a charge, where the "effective force" is given by the expression in square brackets under the integral in (4.4) and includes both the external force and a force on the charge due to its own radiation field. Thus, the acceleration at τ is due to the force at τ *plus* all nonzero forces on the worldline of the charge at all later times, where (due to the strong exponential damping factor in the integral) forces contribute less and less the farther they are in the future. Since, according to (4.4), the acceleration at τ is partly due to effective forces at times other than τ, forces can act where they are not in Dirac's theory. Thus, Dirac's theory violates Lange's condition of spatiotemporal locality. Since a complete set of the causes of a charge's current acceleration includes the field forces acting on the worldline of the charge all the way to future infinity, there is no finite time interval $\Delta\tau$ which contains all of the acceleration's causes.

The nonlocal feature of the theory is perhaps most evident in the following approximation to (4.4). Since τ_0 is small, the equation of motion becomes in first approximation

$$ma^\mu(\tau) = K^\mu(\tau + \xi\tau_0),^{17} \qquad (4.5)$$

where $K^\mu = F^\mu - (m\tau_0/c^2)a^\lambda a_\lambda v^\mu$ is the effective force. Equation (4.5) almost looks like a Newtonian equation of motion, except for the fact that there is a time delay between acceleration and effective force: The acceleration at τ depends on the force at a slightly later proper time. That is, for time differences smaller than $\xi\tau_0$ there is no complete set of causes of the particle's motion. The nonlocal dependence between force and acceleration could be spelled out in terms of what the consequences of counterfactual interventions into an otherwise closed system would be:[18] If we were to introduce an additional external force at $\tau + \xi\tau_0$, the acceleration at τ would have to have been different. By contrast, an intervention into a purely Newtonian system would affect only the acceleration at the time of the intervening force.

4.2. The Role of the Asymptotic Condition

Against the standard interpretation of Dirac's theory, which I am presenting here, Adolf Grünbaum has argued that in Dirac's theory acceleration plays a role analogous to that of velocity in Newtonian theories (Grünbaum 1976). If the analogy would hold, it would be a mistake to interpret (4.4) retro-causally and nonlocally. Grünbaum points out that we can write down equations which allow us to *retrodict* a particle's present Newtonian velocity from its final velocity (as "initial" condition) together with forces acting along the future trajectory of the particle, but this does not show that these forces *retro-cause* the current velocity. For example, the present velocity of a Newtonian particle in terms of its final velocity at future infinity $v(\infty)$ and future forces F is

$$v(t) = v(\infty) - \frac{1}{m}\int_t^\infty F(t')dt'. \qquad (4.6)$$

But obviously we should not interpret this dependence of the velocity at t on future forces causally. Equation (4.6) specifies how we can *retrodict* the present velocity from future data, yet this does not imply that future forces *retro-cause* the present velocity. Similarly, we should not take the fact that (4.4), which is derived with the help of the asymptotic condition as final condition, allows us to retrodict the present acceleration from future forces, to suggest that future forces cause the present acceleration.

There are two different arguments that Grünbaum might be advancing. The first argument appeals purely to the formal similarity between the two equations at issue: Since the Newtonian equation and (4.4) are formally similar, the former ought to be interpreted causally exactly if the latter should be. Yet (4.6) clearly should not be interpreted as specifying a causal relationship between the present velocity and future forces. Thus, the former equation ought not to be interpreted causally either. This argument is intended as a reductio of the retro-causal interpretation of Dirac's theory. If we want to interpret Dirac's theory retro-causally, we are forced to interpret the analogous Newtonian equation retro-causally as well. But this interpretation of the Newtonian case conflicts with the standard interpretation of Newtonian physics and is unacceptable.

Yet this argument—which is in fact not an argument, I think, that Grünbaum would want to endorse—is not very convincing. For it is not clear why we should accept its premise that identical functional dependencies have to represent identical causal structures. Perhaps, however, the formal analogy can support the following weaker argument: Given that (4.4) is formally analogous to the noncausal equation (4.6), interpreting (4.4) causally needs to be justified by an argument for the difference in interpretation. Thus, the objection may be that physicists and philosophers take the formal structure of (4.4) to imply a retro-causal relationship, and that this is mistaken. As we will see below, however, a justification for a retro-causal interpretation of the equation can be given.

The second argument, which appears to be the objection Grünbaum is in fact advancing, points to a purported analogy between (4.4) and (4.6) that goes deeper than the formal analogy—namely to an analogy in the ways in which (4.4) and its purported Newtonian analogue are derived. According to Grünbaum,

> [t]he functional dependence in *Dirac's* theory does *not* have the *law*like character required for qualifying either as asymmetrically causal *or* even as symmetrically causal; instead this functional dependence is crucially predicated on the imposition of a merely *de facto* boundary condition in the form of a constant of integration. (1976, 170; italics in original)

Grünbaum maintains that any causal relationship would have to be underwritten by a purely lawlike equation. However, (4.4), similar to its Newtonian analogue, is not lawlike, since it is derived from (4.3) in conjunction with the asymptotic condition; and this latter condition, according to Grünbaum, is merely a de facto initial condition. In the Newtonian case, the condition that the final velocity is $v(\infty)$ is a purely contingent initial condition, and similarly, Grünbaum maintains, is the condition of zero outgoing acceleration in the case of Dirac's theory. Thus, while the Lorentz–Dirac equation (4.3) is lawlike, equation (4.4) is not.

But Grünbaum appears to misunderstand the role of the asymptotic condition in Dirac's theory, and his analogy is flawed. The asymptotic condition does not merely play the role of a contingent initial condition, as Grünbaum claims, since it needs to be assumed in the derivation of the Lorentz–Dirac equation. Recall that the Lorentz–Dirac equation is derived from the Maxwell equations and the principle of energy–momentum conservation by surrounding the worldline of a charge by a 'tube' and considering the energy balance of the tube as the radius of the tube is shrunk to zero. As we have seen, only if one assumes that the initial and final velocities of the particle are constant, do the infinite self-energy terms have the right form for a mass renormalization to be possible. For only then does the difference between the integrals over the caps of the tube surrounding the worldline of the charge look formally like a difference between particle four-momenta, $m_f u_{out} - m_f u_{in}$, where m_f is the diverging integral that is identified as a 'field mass' of the particle, and u_{out} and u_{in} are the particle's final and initial four-velocities (see Parrott 1987, sec. 4.3). Thus, unlike the Newtonian case, the differential equation of motion (4.3) cannot be derived without assuming the asymptotic condition.

There is, then, a disanalogy between the status of the Newtonian equation of motion and that of the Lorentz–Dirac equation. In the former case, Newton's laws either are taken to be fundamental laws or are derived from a variational principle in Lagrangian or Hamiltonian mechanics. Neither perspective requires us to assume a particular value for the final velocity of a particle. The Lorentz–Dirac equation, by contrast, is generally understood as a derived law. Recall from the previous chapter that the motivation for considering the equation in the first place is that the Newtonian Lorentz force equation of motion is inconsistent with energy conservation and the Maxwell equations for finitely charged discrete particles. Equation (4.3) is derived as an alternative to the Lorentz force law by taking the Maxwell equations and the principle of energy–momentum conservation as basic and considering what form a particle equation of motion would have to take in order to be consistent with these basic principles. Important for our purposes here is that the asymptotic condition crucially enters into the derivation of (4.3), since the expression for the energy balance is not well-defined unless we assume the condition.

To say that the asymptotic condition is *necessary in deriving* the Lorentz–Dirac equation is not to say that it is *a necessary condition* of the Lorentz–Dirac equation. There are solutions to the Lorentz–Dirac equation which violate the condition. Thus, if someone wanted to postulate the Lorentz–Dirac equation as basic equation of a theory irrespective of its relation to the Maxwell equations, then the asymptotic condition would indeed have only the character of a de facto initial condition in that theory. But this is not how physicists think of the equation. The Lorentz–Dirac equation is taken to be part of a theory that also contains the Maxwell equations and energy–momentum conservation and, in fact, is derived from these principles. Such a derivation is impossible without assuming the asymptotic condition.

Now, the asymptotic condition is motivated by assuming that the field with which the charge interacts is localized. But this assumption alone is insufficient for arriving at the asymptotic condition. In addition we need to postulate that the acceleration of a charge vanishes sufficiently far away from any field force. The latter postulate is a weakened principle of inertia: In analogy to the Newtonian principle, it requires that forces be present *somewhere* on the worldline of a charge for the acceleration to be nonzero, but unlike the Newtonian principle, it does not require that the force be nonzero *at the time* of acceleration. This points to an additional problem for Grünbaum's analogy, for Grünbaum claims that acceleration plays a role in Dirac's theory analogous to that of *velocity* in Newton's theory.[19] But there is, of course, no principle analogous to the weakened principle of inertia of Dirac's theory governing velocities in Newtonian physics—a principle saying that velocities go to zero sufficiently far away from any force. If anything, some such principle appears to have been part of Aristotelian physics.[20]

Thus there is a crucial *disanalogy* between velocity in Newton's theory and acceleration in Dirac's—Diracian accelerations vanish far away from any force, but Newtonian velocities do not—while there is an important *analogy* between the roles of acceleration in both theories, despite the formal difference in the equations of motion. In fact, both theories satisfy what Einstein identified as the *law of inertia*: "A body removed sufficiently far from other bodies continues in a state of rest or of uniform motion in a straight line" (Einstein [1916], 1961, 13).

And this analogy is crucial to the causal interpretations of the two theories. For like the Newtonian principle, the principle of inertia in the case of Dirac's theory is quite naturally interpreted causally: The reason why the acceleration of a scattered charge vanishes sufficiently far away from a localized scattering field, is the absence of any cause of the charge's accelerating. Now, of course, a mathematical formulation of the weak principle of inertia does not imply a causal interpretation of the principle. But a causal interpretation is strongly suggested by the similarity between the principle and its Newtonian analogue and the fact that the Newtonian principle is generally interpreted causally—an interpretation that Grünbaum endorses. Moreover, given the weak principle of inertia, Dirac's theory supports certain counterfactuals which strongly suggest a backward-causal interpretation. For the principle of inertia helps to exclude runaway solutions to the Lorentz–Dirac equation and, hence, the theory implies that a charge on which no force is acting at any time moves with constant velocity. But if a force pulse were acting on a charge, then the charge would accelerate prior to the force.

This essentially concludes my reply to Grünbaum. In order to motivate the asymptotic condition, which is necessary for a well-behaved particle–field theory with the Lorentz–Dirac law, the theory is assumed to include a principle of inertia, which 'cries out' to be interpreted causally. Thus, the analogy between Dirac's and Newton's theories is located at a rather different place from the one Grünbaum suggests, and his criticism of the backward-causal interpretation fails. Yet, I think it is useful to examine Grünbaum's claim that the asymptotic condition is a mere de facto constraint in some more detail.

The weakened principle of inertia is presumably best understood as 'lawlike' and is not a contingent initial condition. What about the other assumption needed to arrive at the asymptotic condition—the assumption that the charge is scattered by a localized field? One might think that the assumption that the fields interacting with the charge are localized has the status of a purely contingent initial condition. A consequence of the assumption that fields are localized is that the final acceleration of the charge vanishes. Similar to the case of position and velocity, a zero initial (or final) acceleration, one might say, is just one of infinitely many possible values for the acceleration. But a contingent fact is one that holds for some physically possible worlds but not for all. That is, there ought to be some possible worlds in which the laws of the theory hold that are 'scattering worlds,' and others that are not. Yet since it is impossible to derive an equation of motion for a charged particle from the Maxwell equations without assuming that the particle's acceleration vanishes at infinity, there are no well-defined Maxwellian worlds with charged particles in which the condition does not hold.

Is the scattering assumption physically necessary instead? But this also does not seem to be the right thing to say. Why, for example, should a spatially infinite world with a background field that is constant throughout space be physically impossible? Or why should not a world be possible in which two charges are permanently attached to one another through a spring (so that they never move infinitely far away from one another)? Of course, in one sense one might say that such worlds are physically impossible, for it is hard to imagine that the real world could be anything like that. For in the real world would not the spring between the two charges eventually weaken? But both these imagined models seem to be of the kind that physicists standardly invoke despite their obviously idealized character. At issue here is not whether we think that such models are compatible with all of physics as we know it, but whether they are possible, given purely electromagnetic constraints on the interaction between charges and fields. Now in both cases there will be a field force acting on a charged particle at infinity. Should we think that nonzero forces at infinity are impossible? One reason for thinking that such worlds should be physically impossible would be that these worlds violated the Maxwell equations or the principle of energy–momentum conservation (since these are the laws of the theory). But this is not obviously so. Rather than these worlds violating the Maxwell equations, we simply would not know *how to apply* the Maxwell equations consistently to these worlds, since we would not know what to do with the infinite self-fields associated with the charged particles.

Moreover, physicists themselves do not seem to think that a vanishing acceleration at infinity is physically necessary, for some of the standard applications of the Lorentz–Dirac equation concern models in which the condition is not satisfied. Examples include a charge undergoing uniform acceleration (Rohrlich 1990, sec. 6.11) and a charge on a circular orbit (Rohrlich 1990, sec. 6.15). Thus, even though the asymptotic condition is necessary for deriving the Lorentz–Dirac equation, it is not treated as an assumption that all the systems to which the equation can be applied need to satisfy. Curiously enough, then, there are models governed by the Lorentz–Dirac equation which violate the condition of vanishing acceleration. Of course, according to the standard picture this should be impossible, for these models do not satisfy one of the conditions that needs to be satisfied for the equation to be applicable.

I want to suggest that the assumption is best thought of as a condition that is in a certain sense *mathematically* necessary. It is an assumption without which the Maxwell equations could not be consistently applied to systems of charged particles. As such, the assumption tells us something about the mathematical modeling of the behavior of charged particles, while the connection of the condition to actual systems that may be modeled with its help is rather murky. Take the case of a charged particle in a central field. The relevant idealized mathematical model that can be derived from the Lorentz–Dirac equation does not satisfy the asymptotic condition. Yet Rohrlich argues that *actual* charged particles in a central field, unlike the model, do satisfy the condition, since no actual system has an imposed force doing work over an infinite period of time (Rohrlich 1990, 136). Or consider a particle undergoing uniform acceleration (Rohrlich 1990, sec. 6.11). Rohrlich *first* solves the equation of motion for this case, where the solution violates the asymptotic condition, and *then* argues that due to the asymptotic condition, uniform acceleration is possible only for a finite time interval. In both these examples the models violate a condition that needs to be made in deriving them—that is, in both cases the Lorentz–Dirac equation is first treated as a basic equation, ignoring its relation to the Maxwell equations—but also in both cases the asymptotic condition is not ignored completely; rather, it is invoked in discussing the model-world relation, and is here treated almost like a physically necessary constraint.

There is an important distinction between two types of models that physicists construct with the help of a theory. On the one hand, there are models that are meant to represent real systems; on the other hand, there are models whose job it is to probe and elucidate certain features of the theory, but are never intended to represent any actual phenomenon. The asymptotic condition seems to play a rather surprising role in distinguishing between the two kinds of models. Since the condition needs to be assumed in the derivation of the equation of motion which we use in constructing the models, one would expect that *all* models, no matter what their purpose, satisfy the condition. Instead, what we find is that only some of the models satisfy the condition and that Rohrlich seems to invoke the condition in order to distinguish between mere exploratory 'Tinkertoy' models and those models which might represent actual phenomena.

The asymptotic condition is mathematically necessary, I want to submit, in a manner similar to that in which certain continuity or smoothness assumptions are necessary in certain derivations. For example, in deriving the Lorentz–Dirac equation one also needs to assume that all relevant functions are analytic—that is, infinitely differentiable—a point to which we will return in section 6. This is not, however, straightforwardly treated as an assumption about the world; for the derivation, one needs to assume neither that it is contingent nor that it is *physically* necessary that actual fields be infinitely differentiable. And as in the case of the asymptotic condition, the analyticity requirement need not be satisfied even by all the models of the theory in which physicists are interested. Thus, one stock toy model of the theory is a charge subject to a step-function field force. The analyticity assumption is a mathematical modeling assumption that may or may not receive a physical interpretation, and it is a subtle issue when the condition may be invoked and when not. What may be surprising is that an assumption concerning 'initial conditions' can play a similar role. There really seems to be no room for this kind of condition in the standard picture. The distinction between contingent initial conditions and physically necessary laws appears to be exhaustive: Either all possible worlds allowed by classical point electrodynamics satisfy the asymptotic condition—then the condition is physically necessary, according to the theory—or there are electromagnetically possible worlds that do not satisfy the condition—then the condition is contingent. But the status of the asymptotic condition appears to fit neither.

I have argued that Grünbaum's objection against a causal interpretation of Dirac's theory fails because the asymptotic condition does not have the status of a purely contingent initial condition in Dirac's theory. But there is one important point about which I am in agreement with Grünbaum. The nonlocal and backward-causal interpretation of Dirac's theory follows not *merely* from the fact that (4.4) is a nonlocal equation; rather, it follows from (4.4) in light of the explicitly causal assumption that forces cause accelerations—an assumption that in turn can be motivated as a natural way to account for the asymptotic condition, via the weakened principle of inertia. I agree, then, with Grünbaum that the causal interpretation of the Lorentz–Dirac equation is not implied by the mathematical structure of the theory alone. In particular, I agree with Grünbaum's criticism in (Grünbaum and Janis 1977) of Earman's claim that the causal asymmetry between force and acceleration is due to the following asymmetry of determinism in the theory: "the external force uniquely determines the external acceleration, but not conversely" (Earman 1976, 19). This claim is mistaken, and thus cannot be used to underwrite a causal asymmetry between force and acceleration. For the force at *one instant* does not determine the acceleration at any instant (as can be read off equations (4.3) and (4.4)); and neither does the value of the acceleration at an instant determine the force at any instant. Moreover, even the value of the force over any *finite time interval* is not sufficient to determine the acceleration; only the force over the *entire* future trajectory determines the value of the acceleration now. However, the value of the acceleration over some finite interval, together with the value of the velocity, *does* determine the value of the force, as we can see from (4.3), since (if the trajectory of the charge is sufficiently well-behaved) the value of the derivative of the acceleration at an instant is fixed by the acceleration in any small neighborhood of that instant.

The fact that Dirac's theory is nonlocal might seem surprising, since it is a theory where all interactions between particles are mediated by fields propagating with a finite velocity. Are not field theories local (almost) by definition? This is, for example, suggested in the remarks by Einstein that I quoted in section 2. But we need to be careful here. The electromagnetic field *alone* is local in that the state of the total field (in a given frame at a certain time) is given by the state of the field in all subregions of space. Moreover, disturbances in the field propagate at a finite speed through the field. Nonlocal features arise when we consider how the electromagnetic field interacts with charged particles. The field *produced* by a charge again is locally connected to the charge: According to the Maxwell equations, radiation fields associated with a charge arise at the location of the charge and propagate away from the charge at a finite speed. But the field *affects* charges nonlocally: The acceleration of a charge at an instant is due to the fields on the entire future worldline of the charge. Belot says that it is a "well-entrenched principle that classical fields act by contact rather than at a distance" (1998, 532), and in this he is surely right. But it is a striking (and underappreciated) fact that in microscopic classical electrodynamics this well-entrenched principle is satisfied only by the inconsistent Maxwell–Lorentz theory, and not by Dirac's consistent theory.

4.3. Superluminal Causation

A second sense in which Dirac's theory is nonlocal is that the theory allows for superluminal causal propagation. On the one hand, the present acceleration of a charge is determined by future fields according to (4.4). On the other hand, an accelerated charge produces a *retarded* radiation field which affects the total electromagnetic field along the forward light cone of the charge. The combination of the backward-causal effect of an external field on a charge and the forward-causal influence of a charge on the total field can result in causal propagation between spacelike separated events. If the radiation field due to a charge q_1 at τ_1 is nonzero where its forward light cone intersects the worldline of a charge q_2, then the acceleration of q_2 at τ_2 will be affected by the field due to q_1, even when the two charges are spacelike separated. Again, one could make this point in terms of interventions into an otherwise closed system: If q_1 were accelerated by an external force, then the motion of a spacelike separated charge q_2 would be different from what it is without the intervention. In principle (if τ_0 were not so extremely small) the causal connection between spacelike separated events could be exploited to send superluminal signals. By measuring the acceleration of q_2 an experimenter could find out whether the spacelike separated charge q_1 was accelerated or not, and therefore it should in principle be possible to transmit information superluminally in Dirac's theory.

Dirac's theory is a relativistic theory because it is Lorentz-invariant. Thus, contrary to what is sometimes said, the Lorentz-invariant structure of a theory *alone* does not prohibit the superluminal propagation of causal processes. In a Lorentz-invariant theory the speed of light is an invariant but need not be an upper limit on the propagation of signals. Moreover, as Tim Maudlin (1994) has shown, we can conceive of laws of transmission between spacelike separated events that could not allow us to pick out a particular Lorentz frame as privileged, which would be in violation of the principle of relativity. Maudlin's discussion focuses on the possibility of 'tachionic' signals that directly connect spacelike separated events. Dirac's theory provides us with an alternative mechanism for superluminal signaling: a combination of subluminal forward- and backward-causal processes.[21]

I can think of two objections to my claim that the theory allows superluminal signaling. First, one could argue that since the effects of the acceleration field of q_1 are 'felt' on the entire worldline of q_2 prior to the point where the worldline intersects the future light cone of q_1 at τ_1, one cannot really speak of signaling between spacelike separated points, since no information that was not already available on the worldline of q_2 is transmitted from q_1. But while I take it that signaling implies a causal connection, the converse does not hold. So even if it were impossible to send signals between spacelike separated charges, this does not imply that the two charges cannot causally affect one another. Moreover, we can imagine a scenario in which two experimenters could use two charges to signal to one another. We only need to assume that the experimenter who is to receive the signal at τ_2 has a detection device that is not sensitive enough to detect the influence of $q1$ on q_2 at times prior to τ_2.

Second, one could object to my appeal to counterfactual interventions. Counterfactual interventions are "miracles," and one should not, it seems, draw any consequences concerning a theory's interpretation from what happens if miracles violating the theory's laws occur. We would, for example, not wish to draw any consequences for the causal structure of the theory from a miracle that created additional charges: Even though the effect of the additional charge will be felt instantaneously over spacelike intervals, this does not show that traditional Maxwell–Lorentz electrodynamics is nonlocal. But the intervention in the situation I am considering differs from one creating a charge in that it does not require an *electromagnetic* miracle. For the additional force accelerating the charge could be a nonelectrodynamic force. Since the laws of electrodynamics are not violated in such an intervention, it appears legitimate to appeal to these laws in assessing the effects of the intervention.[22]

Dirac's theory, then, violates two causal locality conditions: the prohibition against action-at-a-distance, as made precise in the condition of *spatiotemporal locality*, and the condition that causal propagation takes place with a finite velocity, that is, *relativistic locality*. The question to which I want to turn next is whether either of these two causal locality conditions can be explicated in noncausal terms.

5. Rigid Body Mechanics and Action-at-a-Distance Electrodynamics

What are the relations between the condition of diachronic locality and the two causal conditions? As we have seen, Belot takes his condition to be equivalent to the principle of finite causal propagation, while Schrödinger proposed an equivalent condition as a noncausal analysis of the prohibition against action-at-a-distance. Putting Belot's and Schrödinger's proposals together, we get the view that the condition that all action is by contact and the condition that causal propagation takes place with a finite velocity are equivalent—a view apparently also endorsed by Healey. And an initial survey of physical theories might seem to support this view. Dirac's theory, as we have seen, violates both conditions, as does the paradigm example of a nonlocal classical theory—Newton's gravitational theory. Pure field theories, by contrast, which are paradigmatic examples of local classical theories, satisfy both conditions.

In fact, however, the two causal conditions are distinct and neither implies the other. On the one hand, spatiotemporal locality does not imply relativistic locality, as the example of nonrelativistic rigid body mechanics shows. In this theory, forces act only by contact, but since extended bodies are treated as rigid, the actions of forces on a body are transmitted instantaneously and infinitely fast throughout the entire body.

On the other hand, action-at-a-distance versions of classical electrodynamics suggest that the converse implication fails as well. In such theories, which are pure particle theories of electrodynamics, forces between distant particles are not mediated by an intervening field and are transmitted across gaps between particles; nevertheless, the force associated with the acceleration of one charge reaches the worldline of another spatially separated charge only after a finite time. Fields are treated in the theory as mere calculational devices (analogously to the treatment of gravitational fields in Newton's theory). We will discuss one example of an action-at-a-distance electrodynamics, the infinite absorber theory of Wheeler and Feynman (1945), in some detail in chapter 6. This theory does not itself provide us with a straightforward counterexample to the claim that the finite propagation condition implies the action-by-contact principle, since the equation of motion for a charge in the Wheeler–Feynman theory is the Lorentz–Dirac equation. Thus, the theory does permit superluminal signaling through the combination of forward-causal and backward-causal effects in the way I have discussed above. Still, the theory suggests how one could 'cook up' a theory that satisfies the finite propagation condition while violating the principle of action-by-contact. For example, a pure particle version of standard Maxwell–Lorentz electrodynamics, which in analogy with the Wheeler–Feynman theory treated fields as mere calculational devices, would be such a theory.

We can appeal to the same two examples to show that neither of the two causal principles implies the condition of diachronic locality. A pure particle version of standard electrodynamics is diachronically nonlocal, even though the effects of one charge on another propagate at a finite speed. Since there is no field that transmits the effects that charges have on one another, they do not 'show up' in a small neighborhood of a test charge before they are felt by that charge. Thus, the present state of the world in a small, finite neighborhood of a charge does not allow us to predict what will happen to the charge next.

Similarly, rigid body mechanics is diachronically nonlocal, since in order to predict, for example, what will happen to one end of a rigid rod next, one always has to look at the entire rod and at whatever other objects are in its immediate vicinity. Thus, the size of the neighborhood of *here* at which we have to look does not shrink to zero as $\Delta t \to 0$. Yet the theory is spatiotemporally local. There is a complete set of causes of the motion of one end of the rod arbitrarily close to that end; and since effects are transmitted simultaneously, the temporal part of the condition is trivially satisfied. Thus the condition of spatiotemporal locality does not imply diachronic locality either.

One might think that, nevertheless, some logical connection between the different conditions exists and that the condition of diachronic locality is strictly stronger than both the condition of action-by-contact and that of finite signaling speeds. But as I want to argue now, the condition of diachronic locality implies neither of the two causal locality principles, since, perhaps surprisingly, Dirac's theory is diachronically local. The only logical connection that might exist between the notion of diachronic locality and the two causal principles is this: The conjunction of spatiotemporal locality and the condition of no superluminal propagation appears to imply diachronic locality. For it is not easy to see how diachronic locality could fail unless either causal influences were transmitted across gaps or causal influences would propagate infinitely fast.

6. Diachronic Locality in Dirac's Theory

It might appear that Dirac's theory is diachronically nonlocal simply because equation (4.4) is nonlocal, since (4.4) appears to indicate that one needs to look at the state of the world at all times later than t in order to determine the acceleration at t. However, this appearance is mistaken for two reasons.

First, (4.4) is not the only way of writing an equation of motion in Dirac's theory. Since in the derivation of (4.4) one needs to assume that the field is an analytic function of the proper time τ, the field and acceleration functions can be expanded in a Taylor series and the nonlocal equation (4.4) is mathematically equivalent to the local equation

$$ma^\mu(\tau) = \sum_{n=0}^{\infty} \tau_0^n \left[F^\mu(\tau) - \frac{1}{c^2} a^\lambda(\tau) a_\lambda(\tau) v^\mu(\tau) \right]^{[n]}, \qquad (4.7)$$

where $[n]$ denotes the nth derivative of the equation in square brackets. Equation (4.7) involves only local quantities, and thus Dirac's theory appears to be diachronically local after all. According to the condition, a theory is local if it is *possible* to determine what will happen *here* by looking at the present state of the world close to *here*, and (4.7) shows that this is possible. But the possibility of representing the motion of a charge in terms of equation (4.7) instead of (4.4) does not affect the causal interpretation of the theory: Even though it might be possible to calculate the acceleration of a charge from the Taylor expansion of the effective force, the force—and not any of its derivatives—is usually taken to be the cause of the acceleration. Thus, (4.7) notwithstanding, Dirac's theory is *causally* nonlocal in the two senses I have distinguished.

One might want to object to this line of argument by claiming that the appeal to analyticity involves some kind of illegitimate trick. The fact that the field is represented by an analytic function, one might say, is merely an artifact of the mathematical formalism and has no physical significance. Thus, the fact that the values of the derivatives of the field *here* allows us to determine the value of the field elsewhere, does not imply that Dirac's theory is diachronically local. A consideration in favor of this response is that physicists treat the analyticity condition rather loosely when they apply the theory. For example, two standard applications of Dirac's theory, which we have already encountered, are the motion of a charge subject to a delta-function field pulse and a step-function pulse, both of which violate the analyticity condition. If the analyticity condition were physically significant, then looking at situations in which the condition is not satisfied and in which, therefore, the theory does not apply is arguably not a good way of investigating what the theory tells us about the world. Moreover, we already know that relying too heavily on the analyticity condition is problematic: Analyticity gives us determinism on the cheap, as Earman (1986, 15) points out. Thus, since diachronic locality is a principle of determinism, it is (not surprisingly) automatically satisfied once we assume analyticity. Thus, the analyticity condition has a status in the theory similar to that of the asymptotic condition: a condition that is *mathematically* necessary but whose representational function is somewhat murky.

Yet even if in the end we want to reject arguments that rely solely on the analyticity condition, doing so without further discussion skirts some important issues. Given that the derivation of Dirac's equation of motion relies on the analyticity condition, why are we allowed to rely on the condition in certain circumstances but not in others? Are there reasons why considerations of analyticity can sometimes be discounted? Why should (4.4) but not (4.7) be a guide to whether Dirac's theory is diachronically local, given that the two equations are mathematically equivalent? I believe that the correct answer to the last question is that (4.4) is privileged in that it represents the causal structure of the theory accurately. If I am right, this means that the condition of diachronic locality could not provide a noncausal explication of either of the two causal conditions, since it would have to be supplemented by the requirement that in predicting what happens *here* next, we have to use an adequate causal representation of the phenomena.

Another question that arises in this context is this: Precisely what quantities characterize the local state of a system *now* and can legitimately be used as inputs to predict future states? Belot's talk of properties which can be looked at is rather vague and is obviously meant only metaphorically, but perhaps one might try to respond to the difficulty raised by the equivalence between (4.4) and (4.7) in the following way. Since derivatives represent *changes* of quantities, derivatives of the field function are not genuinely local quantities, as David Albert (2000) argues, and thus cannot be used as inputs in the condition of diachronic locality. But the problem with this suggestion (aside from the worry that it is far from clear why we cannot think of derivatives as genuinely local quantities) is that we now get nonlocality too easily: Even Newtonian mechanics would come out as diachronically nonlocal, since Newton's laws require velocities, which are derivatives, as inputs.

Dirac's theory comes out as diachronically local for a second reason—one that is independent of the analyticity requirement. Looking at the present state of a system *here* presumably reveals not only the charge's position and velocity but also the local value of the acceleration function.[23] Since disturbances in the electromagnetic field propagate at a finite speed, we can in addition determine the fields on the worldline of the charge during a time interval Δt into the future by determining the fields now in a finite neighborhood of *here*, where this neighborhood shrinks to zero as Δt goes to zero. But then we can use equation (4.3), the Lorentz–Dirac equation, to determine the trajectory of the charge during the time interval Δt. The state of the charge *here*, together with the field in a finite neighborhood of *here*, allows us to predict what will happen *here* in a finite amount of time. This does not conflict with the fact that the theory is causally nonlocal in the two senses I have distinguished, since the effects of future fields—such as that of the field of a signaling charge q_1—are already encoded in the present acceleration of the charge q_2 here. Thus, even though we can determine the local evolution of the system from local data, this does not imply that what happens to the system is due only to locally acting causes whose effects propagate at a finite speed.

7. Localizations of Global Models

I have argued that the condition of diachronic locality is conceptually distinct from both the condition that causal propagation occurs at a finite speed and the condition that all action is by contact. Diachronic locality is what one might call *a condition of local determinism*: We can predict what will happen *here* next if and only if the evolution of a localized subsystem is completely determined by the local state of the system—the state of the system in a finite neighborhood of *here*. For if the evolution of a local subsystem is not completely determined by the local state, we cannot know what will happen *here* next without looking elsewhere; and similarly, if the evolution of the subsystem is determined by the local state, then we can use our knowledge of the local state to predict what will happen *here* next.[24]

Earlier I suggested that the conjunction of spatiotemporal locality and the condition of no superluminal causal propagation might imply diachronic locality, though we have seen that neither of the two causal principles separately implies the latter principle. Thus one might think that the problem with diachronic locality as putative explication of the two causal principles is that the condition is not sensitive to the distinction between the two causal principles. Perhaps, then, the project of noncausally explicating the two causal principles need not fail in general. Might there be other noncausal principles that could individually explicate the causal principles and are sensitive to differences between the two causal principles?

As far as the prohibition against superluminal causal propagation is concerned, I think the answer is 'no.' I can think of only two plausible kinds of candidates for such an explication. The first would appeal to special relativity and the Lorentz–invariant structure of space-time, and the second would invoke the fact that the Maxwell equations imply a hyperbolic partial differential equation for the field coupled to a charged particle, according to which field disturbances propagate at a finite speed. But Dirac's theory, of course, is a relativistic Maxwellian theory and nevertheless allows for superluminal causal propagation. To be sure, the superluminal connection is due to the combination of a subluminal forward-causal and a subluminal backward-causal process. Yet it is not easy to see how one could express a prohibition against this kind of combination entirely in noncausal terms.

The prospects are somewhat better as far as the prohibition against unmediated causal interactions is concerned. Earman has proposed the following condition as a possible explication of the action-by-contact principle: "Every localization of a global model of T is again a model of T" (Earman 1987, 455), where a localization is a restriction of a model of T to a neighborhood U which is a subset of the manifold M on which the models are defined.[25] If a localization of a global model is again a model of the theory, then it does not matter to local properties of the system whether these properties can 'see' the values of quantities far away. That is, if a localization is itself a model of the theory, then all properties of a local subsystem are completely determined by the values of local quantities. Thus Earman's condition, like diachronic locality, is a condition of local determinism. Yet Earman's condition is distinct from the principle of diachronic locality, as the example of nonrelativistic rigid body mechanics shows: That theory is, as we have seen, diachronically nonlocal, yet it satisfies Earman's condition.

Does Dirac's theory satisfy Earman's condition? That depends, once more, on which equation we take to be the fundamental equation of motion of the theory. If the nonlocal equation (4.4) is taken to be the fundamental equation of motion, then Dirac's theory does not seem to obey Earman's condition. According to (4.4), the acceleration of a charge that 'sees' only the fields in a finite neighborhood U of the charge should in general be different from the acceleration in a global model (and hence in a localization of the global model), in which the acceleration also depends on future fields outside of U. In fact, however, it is not completely obvious how one should apply Earman's criterion to theories with 'global' equations of motion like (4.4). One cannot simply restrict (4.4) by changing the limits of integration, but if one 'localizes' the fields by multiplying them by a step function, the resulting field function no longer is analytic. Nevertheless, since in order to assess what (4.4) predicts for the acceleration we need to be given the fields *everywhere* on the future worldline, I am assuming that the restriction of the fields to a finite neighborhood U could be obtained by multiplying the field function by a step function whose value is 1 inside of U and zero outside.

In any case, if we take (4.4) as a guide, then it seems that a localization of a global model will not in general be a model of Dirac's theory. If, however, the local equation (4.3) or (4.7) were to be taken to be the fundamental equation of motion, then the theory would appear to satisfy Earman's condition. Since, according to (4.7), the acceleration is determined from the local values of all the derivatives of the field, a localization of (4.7) is itself a model of (4.7). Dirac's theory would come out as local, according to Earman's condition, even though it violates the condition of spatiotemporal locality. Now, I have already suggested that one way to decide which of the two equations is fundamental is to appeal to the causal structure of the theory. Equation (4.4) is more fundamental, since it gives the acceleration in terms of its causes, that is, the electromagnetic fields. But if Earman's condition needs to be supplemented by considerations concerning the causal structure of the theory, then it cannot provide us with a strictly noncausal explication of the action-by-contact principle.

8. A Russellian Objection

The failure of diachronic locality and of Earman's principles of determination to provide adequate explications of the two causal locality conditions *spatiotemporal* and *relativistic locality* suggests that there is more to causation than determination. And this in turn implies that in the case of scientific theories that involve causal claims, the job of an interpretation cannot be exhausted by stipulating how the mathematical formalism maps onto the ontology of the theory. For a theory's causal structure will generally be constrained by the mathematical formalism but not uniquely determined by it. In classical electrodynamics it is part of the causal structure that force (and not any of its derivatives) is the cause of acceleration; but this cannot be read off the theory's formalism alone. Whether a theory is causally nonlocal in either of the two senses I have distinguished depends crucially on the causal interpretation of the theory. Yet whether a theory satisfies a condition of local determinism depends only on the theory's mathematical formalism (and the associated ontology).

At this point the following response suggests itself: The fact that the intuitive causal claims associated with a scientific theory can outrun what can be legitimately inferred from the theory's mathematical formalism, and the fact that taking the causal locality principles too seriously can lead to rather strange and counterintuitive results (as in the case of the putatively backward-causal theory of Dirac), only further support Russell's view that a rich notion of causality that cannot be reduced to that of functional dependency should have no place in science. Einstein's notion of locally transmitted influences cannot be spelled out in terms of a principle of determinism, but so much the worse for the principle of *local action*.

I want to make two remarks in response to this objection. The first is that a theory's causal interpretation can play a significant methodological role. In the case of Dirac's theory (as we have seen above), causal assumptions play an important role in motivating various steps in the derivation of the theory's equation of motion. First, the assumption that field forces are the cause of a charge's acceleration makes plausible the adoption of what I have called *the weak principle of inertia*, according to which the acceleration of the charge should vanish far away from any forces. This principle then helps to motivate the asymptotic condition of vanishing accelerations at infinity. Without the causal framework, the asymptotic condition can be given only a purely mathematical motivation: One can renormalize the mass only if the condition is presupposed. Assuming that fields cause charges to accelerate provides a physical reason for the condition as well. And second, the causal assumption helps to motivate the rejection of runaway solutions to the Lorentz–Dirac equation as unphysical. For given that fields cause a charge to accelerate, a charge should not accelerate in the absence of any external fields anywhere.

My second remark is that a causal interpretation of Dirac's theory also seems to be at the root of the feeling of unease that many physicists have toward the theory. For the causal structure of the theory violates several requirements we would like to place on causal theories. If nothing more were at issue than questions of determination, the nonlocal character of the equation of motion ought not to be troubling. That is, physicists themselves appear to be guided by causal considerations in their assessment of the theory. Now of course it is always open to a Russellian to reject this interpretation of the theory by physicists as misguided. But it is important to be clear on what the form of the argument against causal notions in fundamental physics then is. The argument cannot rely on the practice of theorizing in physics, for that practice does involve appeals to causal notions, even if the precise content of such appeals is not always spelled out carefully. Rather, the argument has to appeal to a certain *philosophical reconstruction* of this practice. But then the question becomes what the support for that particular kind of reconstruction is and why it is superior to accounts that allow for weighty causal notions. The way physicists treat Dirac's theory of the electron shows that Russell was wrong: Weighty causal notions are part of theorizing in fundamental physics.

One may wish to maintain that such notions ought not to play a role in physics. But this is a normative appeal to what the proper content of science should be. Contrary to what Russell himself suggests, the prohibition against causal notions receives no immediate support from a description of scientific practice.

9. Conclusion

In this chapter I distinguished several different locality conditions and argued that two of these principles are irreducibly causal—the principle that all causal propagation takes place at a finite velocity, of which *relativistic locality* is a special case, and *spatiotemporal locality*, according to which no causal influences can be transmitted across spatiotemporal 'gaps.' In particular, I have argued that neither of the two causal principle can be reduced to the noncausal condition of diachronic locality. Since any noncausal explication of the two prima facie causal principles apparently would have to invoke some principle of local determinism, and since the concept of causation does not seem to be reducible to that of determinism, the prospects for a successful empiricist reduction of either of the two causal locality principles appear to be dim.

PART TWO

FIELDS

5

The Arrow of Radiation

1. Introduction

When a stone is dropped into a pond whose surface was still initially, we observe waves diverging circularly from the point of impact. The time reverse of this phenomenon—circularly converging waves—does not seem to occur in nature. In electrodynamics this 'wave asymmetry' is exhibited by radiation fields. The coherently diverging radiation fields associated with an accelerated charge are the temporal inverse of coherently converging fields (imagine a film that depicts a diverging wave run backward), but only the former fields seem to occur in nature. This apparent asymmetry might strike one as particularly puzzling, given that the laws we use to describe these phenomena are invariant under time reversal. The Maxwell equations are (in a sense to be explained shortly) time-symmetric, and allow for both converging and diverging fields to be associated with sources of radiation. If the underlying laws are time-symmetric, then where does the temporal asymmetry come from? This is the problem of the arrow of radiation.

In the remaining chapters of this book I will examine several accounts of the asymmetry. This chapter will largely be devoted to setting the mathematical and historical stages for the discussion in subsequent chapters. One important issue will be to get clearer on precisely what the puzzle is that might be in need of an explanation. We will see that the intuitive characterization of the problem I just gave, while useful as an introduction to the problem, is in fact not fully adequate and will have to be refined.

Contemporary discussions of the radiation asymmetry sometimes introduce the topic by referring to a debate between Albert Einstein and Walter Ritz (Einstein 1909b; Ritz and Einstein 1909) and to Karl Popper's brief discussions in a series of letters to the journal *Nature* of waves spreading on a surface of water (Popper 1956a, 1956b, 1957, 1958). After presenting some of the necessary background, I will follow this precedent but will try to correct certain misrepresentations of Popper's and Einstein's views in the literature. I will conclude this chapter by briefly discussing two more recent attempts to solve the puzzle offered by James Anderson (1992) and Fritz Rohrlich (1998, 1999, 2000), which, unlike the various accounts I will discuss in the next chapter, do not invoke thermodynamic considerations to account for the asymmetry. While both accounts will allow us to get certain aspects of the puzzle into clearer focus, neither of the two proposals, I will argue, offers a satisfactory solution.

Even though I will focus on the arrow of radiation, much of my discussion should generalize to other instances of the wave asymmetry, such as that of waves spreading on a pond, since the mathematical formalism in all these cases is the same. The asymmetry of radiation and of waves more generally is only one of a number of temporal asymmetries in nature. Other examples are the thermodynamic asymmetry according to which the entropy of a closed system never decreases, the cosmological asymmetry of an expanding universe, and the causal asymmetry that causes universally seem to precede their effects. Many discussions of the radiation asymmetry seem to be motivated by the desire to find connections between the various 'arrows' and by attempts to reduce them to a single 'master arrow.' If it were possible to reduce all temporal asymmetries to a single master arrow, one might even try to argue that this arrow not only is an asymmetry *in* time but also is constitutive of the arrow *of* time—that is, that this master arrow constitutes the distinction between past and future. This master arrow then also ought to account for our psychological perception of time as in some sense flowing from the past into the future. I think however, that any such grand unifying project should be met with a healthy dose of skepticism.

The aim of this and the following chapters will be to get a clearer understanding of the relations between the arrow of radiation and other temporal arrows. In chapter 6, I will examine several accounts of the arrow of radiation that have their roots in the infinite absorber theory of John Wheeler and Richard Feynman, and aim to reduce the radiation asymmetry either to the thermodynamic asymmetry or to a cosmological arrow. As I will argue, all such absorber theories of radiation ultimately fail to provide a satisfactory solution to the puzzle of the arrow of radiation. There is, however, a grain of truth in absorber theories of radiation: Thermodynamic considerations do play a role in accounting for the asymmetry of radiation. Yet, as I will argue in chapter 7, a successful explanation of the asymmetry needs to appeal to a causal asymmetry as well, which in the end does most of the explanatory work.

2. What Is the Asymmetry of Radiation?

The fundamental equations governing radiative phenomena are the Maxwell equations.[1] Strictly speaking, the Maxwell equations are *not* time-reversal-invariant—that is, the equations are not invariant under the transformation $t \to -t$.[2] But if we add to the time reversal transformation a space reflection, as a second compensatory transformation, then the Maxwell equations are invariant under the combined transformation. The effect of the space reflection is to invert the sign of the magnetic field vector. As Dieter Zeh explains it:

All known *fundamental* laws of nature are symmetric under time reversal after compensation by an appropriate symmetry transformation, T, [... where] $T\{\mathbf{E}(\mathbf{r}); \mathbf{B}(\mathbf{r})\} = \{\mathbf{E}(\mathbf{r}); -\mathbf{B}(\mathbf{r})\}$ in classical electrodynamics. This means that for any trajectory $z(t)$ that is a solution of the dynamical laws there is a time-reversed solution $z_T(-t)$, where z_T is the "time-reversed state" of z, obtained by applying the compensating symmetry transformation. (Zeh 1989, 4)

Generally, then, when people speak of 'time reversal' in classical electrodynamics, they refer to the combined transformation. Independently of whether one thinks that it is appropriate to call the combined transformation the 'time-reversal operation,' what appears to be puzzling about radiative phenomena is that they exhibit a temporal asymmetry *even if* we include a compensatory transformation on the magnetic field.

In first introducing the asymmetry of radiation, I suggested that the puzzle was given by the fact that there are diverging fields associated with charged particles, or wave sources, but not with converging waves. And this is in fact how the asymmetry is often characterized in the literature. While this characterization can be useful in getting a preliminary understanding of the asymmetry, it also is misleading in important respects. For in an important sense every radiation field in the presence of charges can be thought of *both* as diverging *and* as converging. Thus, I now want to work toward a more precise formulation of what the puzzle of the arrow of radiation is.

Radiation phenomena are modeled by assuming that particle trajectories are given independently and by using the Maxwell equations to determine the effect of charged particles on the electromagnetic field.[3] The radiation fields are usually calculated by introducing the electromagnetic four-potential A^μ, which is related to the field tensor via $F^{\mu\nu} = \partial^\mu A^\nu - \partial^\nu A^\mu$, and then adopting the Lorenz gauge $\partial_\mu A^\mu = 0$. The potential then satisfies the wave equation

$$\partial^\nu \partial_\nu A^\mu(\mathbf{r}, t) = 4\pi j^\mu(\mathbf{r}, t). \tag{5.1}$$

This equation is an *inhomogeneous* partial differential equation, which means that the right-hand side of the equation specifying the distribution of the electromagnetic charges, or sources, to which the field is coupled, is nonzero. If the source term is zero, then the equation is what is known as a *homogeneous* partial differential equation, solutions to which represent source-free fields. One can obtain different solutions to the inhomogeneous equation by adding solutions to the homogeneous equation to any one solution to the inhomogeneous equation. Differential equations are equations that specify how the values of certain variables change. In order to obtain a particular solution to a differential equation, one needs to specify 'what the values change from,' that is, one has to specify particular initial conditions. The homogeneous equation is generally solved in terms of a pure initial (or final)-value problem—a *Cauchy problem*. In the case of an infinite spatial volume, the initial values are given by the value of the potential and its derivatives on a spacelike hypersurface. It is an important fact that any solution to the inhomogeneous equation matching particular boundary conditions can be expressed in terms of any other solution to the inhomogeneous equation, if an appropriate solution to the homogeneous equation is added.[4]

The problem of finding the field associated with an arbitrary source distribution is usually approached in terms of *Green's functions*, which specify the field component associated with a point charge. The total field associated with a given source distribution can be determined by integrating over all the infinitesimal contributions to that source distribution. One particular solution to the wave equation for the electromagnetic potential of a single point charge specifies the potential at a field point P in terms of the unique space-time point Q at which the worldline of the charge intersects the past light cone of P. This solution is known as the *retarded solution*. If we focus on the point Q on the worldline of the charge that is picked out by the retarded solution, then all field points whose associated retarded potentials pick out Q lie on the future light cone of Q. Thus, at later and later times, field points farther and farther from Q depend on the charge at Q; that is, the retarded solution represents an electromagnetic disturbance concentrically diverging from Q into the future. Another solution specifies the potential in terms of the unique intersection of the worldline of the charge with the future light cone of the field point P. This is the *advanced solution*. The advanced solution represents an electromagnetic disturbance concentrically converging into a source point Q' from the past.

Since any solution to the wave equation can be represented as the sum of an arbitrary solution to the inhomogeneous equation and free fields, any solution can be represented as the sum of a retarded field and a free (incoming) field, $F = F_{ret} + F_{in}$ or, similarly, as the sum of an advanced field and a free (outgoing) field, $F = F_{adv} + F_{out}$. Moreover, since the wave equation is linear, any linear superposition of solutions will also be a solution. Thus, the most general solution to the wave equation can be written as

$$F = k(F_{ret} + F_{in}) + (1-k)(F_{adv} + F_{out}). \qquad (5.2)$$

The free field component of the retarded solution is called an *incoming field*, since, in the retarded case, the free field contribution to the value of the field at P in a certain region R of space-time is given in terms of the value of the free field at the intersection of the *past* light cone of P with a spacelike hyperplane which constitutes the past boundary of R. The retarded solution is the solution to an *initial-value problem*. At the past boundary $F = F_{in}$, and F_{ret} specifies how the field changes due to the presence of the charges. Similarly, the free field is called *outgoing* in the case of the advanced solution, because the relevant boundary conditions are now those on a hyperplane in the *future* of P—that is, the field is now given as a solution to a final-value problem. Both the initial- and the final-value representations are representations of one and the same field, and both are mathematically correct representations of the field. Notice that retarded and advanced field problems are not pure Cauchy problems. While the incoming or outgoing fields are specified on an initial-value surface, the retarded or advanced fields depend on the *entire* trajectory of the charged particles in question.

So far there is no asymmetry. *Every* electromagnetic field can be represented both as a retarded and as an advanced field, when appropriate free fields are added, depending on whether the wave equation is solved as an initial-value or a final-value problem. In what sense, then, is electromagnetic radiation asymmetric? The temporally asymmetric, diverging fields we observe are very nearly fully retarded fields. That is, in these situations it appears possible to choose boundary conditions such that the free incoming field F'_{in} is nearly equal to zero in a retarded representation of the total field. Of course, a very nearly fully retarded field can alternatively be represented as the sum of an advanced field and a source-free field, $F_{adv} + F_{out}$, but the two representations are not symmetric: The latter representation will in general include a nonnegligible source-free field, while the former does not.[5] Thus, the asymmetry associated with radiation phenomena is an asymmetry between initial and final conditions.[6] And the puzzle of the arrow of radiation might then be put as follows:

1. Why does the Sommerfeld radiation condition $A^\mu_{in} = 0$ (in contrast to $A^\mu_{out} = 0$) approximately apply in most situations?
2. Why are initial conditions more useful than final conditions? (Zeh 2001)

Zeh's second question can be answered by appealing to the presuppositions of the first: Initial conditions are more useful than final conditions because zero incoming fields are mathematically more tractable than arbitrarily complex nonzero outgoing fields. But while Zeh's first question points in the right direction, it is still not an entirely adequate characterization of the problem. I want to discuss four different worries about Zeh's formulation of the puzzle. Some of the worries can be diffused, but others make a reformulation of the problem necessary.

First, one might object that incoming fields are rarely, if ever, strictly zero. Even in circumstances where most other incoming radiation is blocked out, the 4 Kelvin cosmic background radiation will still be present.[7] Also, since charge is conserved, charges associated with the radiation fields must have existed at times before the initial time, and the field on the initial-value surface should at least include the Coulomb field associated with these charges. However, the fact that incoming fields may not be strictly zero is already accounted for in Zeh's formulation of the puzzle, and we can stipulate that pure Coulomb fields ought to be ignored.

Second, the asymmetry seems to manifest itself even in circumstances where incoming fields are noticeably different from zero—for example, if a light is turned on in a room that is not completely dark. Thus, there also seems to be an asymmetry in situations not captured by Zeh's formulation. This worry can be met if we construe the formulation not as offering a comprehensive account of *all* phenomena exhibiting the asymmetry, but rather as delineating one relatively precisely characterized and central class of phenomena exhibiting the asymmetry. That is, we assume that there is a general asymmetry governing radiation phenomena, but begin our search for an account of that asymmetry by delineating a class of phenomena for which the asymmetry is particularly pronounced and relatively easy to characterize. The task in trying to find an explanation of the asymmetry for this class is then to find an account that is general enough to extend to other, perhaps only more vaguely characterizable, cases as well.

Third, even in those situations for which initial-value surfaces with approximately zero incoming fields can be found, there are also many spacelike hypersurfaces on which the fields are nonzero that could serve as initial-value surfaces as well, but that do not result in purely retarded fields. For example, we can model a radiating charge by picking an initial-value surface at some time *after* the charge has begun radiating. In what sense, then, does the condition $A^\mu_{in} = 0$ "apply in most situations"? And, fourth, there seem to be just as many situations $A^\mu_{out} = 0$ as there are situations $A^\mu_{in} = 0$, for yet another reason: Every spacelike hypersurface can serve *both* as an initial-value surface for the fields in its future *and* as a final-value surface for fields in its past. But then it might seem that radiation is not asymmetric after all. While both initial-value and final-value representations are individually asymmetric, radiation *as a whole* might seem to be symmetric, since every asymmetric representation with zero incoming fields is balanced by one with zero outgoing fields.

We can address the final worry once we realize that there is an additional asymmetry between situations with zero incoming and zero outgoing fields. Many phenomena for which the total field is approximately equal to a sum of retarded fields involve only the retarded fields associated with a small number of sources. By contrast, in situations where the total field is very nearly fully advanced, the total field is (almost) always a sum of advanced fields of a very large number of sources. Roughly — and we will have the opportunity to discuss this in much more detail in later chapters — the electromagnetic field can be zero in a certain region of space at a certain time despite the fact that there were coherently radiating sources in the region's distant past, if there is an absorbing medium in the region's more recent past. And absorbing media are usually modeled as consisting of a very large number of charges that scatter and damp out any incoming field. Thus, regions with approximately zero fields will have either no charged particles in their (recent) past or a very large number. By contrast, such regions may have a small number of charges in their future.

Finally, if we express the puzzle as postulating the existence of spacelike hypersurfaces on which the incoming fields are zero, then the third objection above can be met as well, and the asymmetry of radiation can now be expressed somewhat more precisely, as follows:

> (RADASYM) There are many situations in which the total field can be represented as being approximately equal to the sum of the retarded fields associated with a small number of charges (but not as the sum of the advanced fields associated with these charges), and there are almost no situations in which the total field can be represented as being approximately equal to the sum of the advanced fields associated with a small number of charges.

The asymmetry captured by RADASYM is an asymmetry between initial-value and final-value representations, and not an asymmetry in the dynamical evolutions of electromagnetic systems. What needs to be explained in answering the problem of the arrow of radiation is why certain initial-value representations are possible but not the corresponding final-value representations.

There is a second putative asymmetry associated with radiation phenomena that is frequently discussed in the literature and that involves the notion of the field physically associated with a charge. This second asymmetry consists in the supposed fact that the field associated with a charged particle, or the field contributed by each charged particle, is fully retarded. This fact may then be used to explain why the asymmetry between initial and final conditions holds: Since electromagnetic fields eventually dissipate and are absorbed by media through which they propagate, and since each source contributes a fully retarded field to the total field, overall, diverging waves are overwhelmingly more probable than converging waves (which would require carefully set-up correlations between many different radiating sources).

One might think that there is a second puzzle of the arrow of radiation corresponding to this second putative asymmetry: Given that the Maxwell equations are symmetric in time and do not by themselves distinguish in any way between retarded and advanced solutions, why is the radiation field physically associated with a charged particle fully retarded (and not fully advanced)? Textbooks in electrodynamics, such as those by Jackson (1975) and Landau and Lifshitz (1951), offer answers like the following to this question: Electromagnetic disturbances propagate at a finite velocity (which in a vacuum is the speed of light c) into the future. Thus, one should expect the field at a time t some distance from an electric charge to depend not on the motion of the charge at t, but rather on the motion of the charge at a time t_R *earlier* than t and at a position R which will in general be different from the charge's position at t, where t_R is determined by the time it takes for the disturbance to travel from the retarded point R to the observation point. The temporal direction in which electromagnetic disturbances propagate, therefore, imposes a "causal" (Jackson 1975, 245; Rohrlich 1965, 77) or "physical" (Rohrlich 1965, 79) constraint on possible solutions to the wave equation, a constraint which is satisfied only by the retarded solution.

In the end I will endorse a causal account of the asymmetry of radiation. Of course this presupposes that we can make sense of the idea of a field being physically associated with a charge. I will defend the legitimacy of this notion in chapter 7. For now, I just want to point out that the notion does not in any obvious way conflict with the facts about differential equations that we rehearsed above. For the fully retarded field associated with a charge can of course also be represented as a partly advanced and partly free field. But if each charge *physically* contributes a fully retarded field to the total field, then the physical interpretations of the *mathematical* initial-value and final-value problems are not identical: In the case of an initial-value problem, the retarded contribution to the total field does directly represent the physical contribution of the charges in the problem. In a final-value problem, however, the advanced field mathematically associated with the charges does *not* correctly give the physical contribution of the motion of the charges to the field. That contribution still would be given by the fully retarded fields, which in a final-value problem are represented as a combination of advanced and free fields.

3. Einstein, Ritz, and Popper's Pond

Contemporary accounts of the radiation asymmetry sometimes cite a debate between Ritz and Einstein in the first decade of the twentieth century as an important precursor to the contemporary discussion of whether the arrow of radiation can be reduced to statistical or thermodynamical considerations (see, e.g., Zeh 2001; Price 1996). Ritz argued that the thermodynamic arrow is reducible to that of electrodynamics, while Einstein is usually taken to have argued in favor of a reduction of the electrodynamic arrow to that of thermodynamics. After a series of papers, Ritz and Einstein summed up their disagreement in a short joint letter (Ritz and Einstein 1909). Both agreed that if we restrict ourselves to finite regions of space, we can represent the same field in terms of the retarded fields associated with the charges in the region and in terms of the advanced fields associated with the charges (if, one should add, one takes into account appropriate free fields at the boundaries). Ritz thought, however, that the theory has to account for the case of infinite spaces as well. Ritz and Einstein wrote:

> If one takes that view, then experience requires one to regard the representation by means of the retarded potentials as the only possible one, provided one is inclined to assume that the fact of the irreversibility of radiation processes has to be present in the laws of nature. Ritz considers the restriction to the form of the retarded potentials as one of the roots of the Second Law, while Einstein believes that the irreversibility is exclusively based on reasons of probability. (Ritz and Einstein 1909, 324)[8]

The difference between the finite case and the infinite case is this. In the case of finite regions it is easy to choose initial-value or final-value value surfaces such that the incoming or outgoing fields will be different from zero. If we assume that all fields are ultimately associated with charged particles, then on a view such as Ritz's, nonzero *incoming* fields will be due to charged particles in the past of that region. But if we consider an initial-value surface in the infinite past, then, both seem to agree, the incoming field has to be zero.

This passage is commonly taken to show that Einstein believed that the asymmetry of radiation can be fully accounted for by thermodynamic and statistical considerations. Thus, Wheeler and Feynman refer to the exchange between Ritz and Einstein in their paper on the infinite absorber theory and align their own view, according to which the asymmetry can be accounted for statistically, with that of Einstein (Wheeler and Feynman 1945, 160). Huw Price maintains that Einstein's view was that the asymmetry of radiation is "of the same origin as the thermodynamic asymmetry" (Price 1996, 50), and Zeh claims that Einstein "favored the point of view that retardation of radiation can be explained by thermodynamical arguments" (Zeh 2001, 16). Like Wheeler and Feynman, Price and Zeh take their own views to be in rough agreement with that of Einstein, despite the fact that there are a crucial difference among all three accounts: In particular, while Zeh believes that the radiative asymmetry can be explained in terms of the thermodynamic arrow (see Zeh 2001, 22), Price thinks that neither arrow can explain the other, but that both have a common source (see Price 1996, 52).

It appears, however, that Einstein's view was considerably more complex than these interpretations suggest. At the very least, I want to argue, Einstein was rather ambivalent in his support for the view that the arrow of radiation can be fully explained in terms of thermodynamic or statistical considerations alone. Quite plausibly he did not at all believe that the asymmetry of "elementary" radiation processes can be explained in this manner. Thus, in what follows, I want to offer a rival interpretation of Einstein's view. The discussion will also provide a useful preview of some of the themes discussed in much more detail in the next chapter.

In a paper preceding the joint letter with Ritz, Einstein explained his reasons for being dissatisfied with Ritz's use of retarded potentials (Einstein 1909a). He said that he thought of retarded field functions "only as mathematical auxiliary devices" (1909a, 185).[9] In the paper Einstein was working with an action-at-a-distance conception of classical electrodynamics (as was Ritz), in which the Maxwell equations in empty space are only "intermediary constructs" (ibid.), and do not describe the state of an independently existing electromagnetic field. Einstein's main complaint against retarded potentials in this context was that they do not provide us with a way to write down the energy of a system at a single instant: "In a theory operating with retarded forces the state of a system at an instant cannot be described without making use of previous states of the system" (ibid.). The problem, according to Einstein, is that in a retarded field representation, radiation cannot be represented at all between the moments when it is emitted by one charge and when it is absorbed by a second charge. Thus, "energy and momentum have to be represented as temporal integrals—if one does not want to give up these quantities altogether" (ibid.).[10] Einstein's concern is a version of the worry that we discussed in chapters 1 and 3—that energy–momentum is not conserved in an action-at-a-distance electrodynamics.

Einstein went on to argue that retarded and advanced representations can be used equivalently, as well as mixed representations of the form $kF_{ret} + (1-k)F_{adv}$. There are no incoming or outgoing free fields in this representation, if we assume that all the radiation emitted by charges is eventually absorbed by other charges. The complete symmetry between the two representations is broken, as Einstein pointed out, if not all the radiation is eventually absorbed and there is a nonzero field "emitted into the infinite" (1909a, 186). One might try to argue, then, that the fact that there can be free outgoing fields, but that any incoming field has to be due to some charge somewhere, leads to a genuine asymmetry and a preference for retarded field representations. But Einstein objected to this argument on two grounds:

> For one, we cannot speak of the infinite, if we want to stay within the realm of experience, but only of spaces that lie outside of the space under consideration. Moreover we cannot conclude from the irreversibility of such a process that electromagnetic elementary processes are irreversible, just as we cannot conclude that the elementary motions of atoms are irreversible on the basis of the second law of thermodynamics. (1909a, 186)[11]

Einstein's second argument might seem to support Price's reading: The elementary processes responsible for electromagnetic radiation are symmetric, Einstein seems to have believed, and there is a strong suggestion that the source of the appearance of irreversibility is related to the source of the thermodynamic asymmetry.

The trouble is that in the very same year and in the very same journal, Einstein also had this to say about the processes governing wave phenomena:

> The basic property of wave theory, which leads to these problems [in trying to account classically for what we today think of as quantum phenomena], appears to be the following. While in kinetic molecular theory the inverse process exists for every process involving only a few elementary particles, e.g., for every elementary collision, according to wave theory this is not the case for elementary radiation processes. An oscillating ion produces a diverging spherical wave, according to the standard theory. The reverse process does not exist as elementary process. A converging spherical wave is mathematically possible; but in order to realize such a wave approximately a tremendous number of elementary objects are needed. *The elementary process of the emission of light is, thus, not reversible.* In this respect, I believe, our wave theory goes wrong. (Einstein 1909b, 819; my italics)[12]

Einstein went on to discuss the advantages of a quantum treatment of radiation that allows for the entire energy radiated by a source to be absorbed in a single "elementary process," instead of being dispersed, as in the wave theory.

Einstein here seems to directly contradict what he said earlier. In the joint letter with Ritz he expressed the view that that the irreversibility of radiation processes "is exclusively based on reasons of probability," while here he maintains that elementary radiation processes are irreversible. Unlike the case of kinetic theory and the second law of thermodynamics, even the fundamental microprocesses of radiation are asymmetric in the wave theory of light, according to Einstein. To be sure, he thought that this aspect of the classical theory of radiation is problematic. But this does not affect the point that here he seems to assert what he apparently denied elsewhere: that *within classical electrodynamics*, elementary radiation processes are asymmetric.

No matter how we try to make sense of this apparent conflict, it is obvious that Price's and Zeh's accounts cannot be the whole story. How do we incorporate the last passage into a coherent interpretation of Einstein's views? I am not sure that there is an easy interpretation of these passages that can render them entirely coherent. Of course, one possibility would be that Einstein simply had changed his mind. While his view at the beginning of 1909 might have been the view attributed to him by Zeh or Price, he might, by the end of that year, have come to accept a view similar to that of Ritz. But I believe there is a view that can accommodate most of what Einstein wrote in 1909 about the classical arrow of radiation. While I do not wish to claim that this is the view Einstein in fact held, the view has certain virtues: It can make much better sense of the totality of Einstein's comments than either Zeh's or Price's account does; and it is a view that is close to the view that, at the end of the day, appears to be the most tenable view overall. (Thus, I will follow Price, Zeh, and Wheeler and Feynman in wanting to appeal, at least sotto voce, to Einstein's authority in support of a certain account of the radiative asymmetry.)

Notice, first, that Einstein's main criticism of Ritz in (Einstein 1909a) is that a retarded formulation of an action-at-a-distance theory does not allow one to write down the energy of a system, and Einstein was not willing to give up energy conservation easily. But this criticism applies to an advanced field representation as well. As we have seen, he maintained that a retarded representation of the total field force should be understood as only a mathematical construct. He presumably thought the same about purely advanced representation in action-at-a-distance theories.[13] Thus, Einstein's dissatisfaction with a retarded field representation seems to have arisen to a considerable extent from worries that are independent of the issue of temporal asymmetries.

Einstein then argued, in essence, that one cannot infer from the fact that under certain circumstances the total field can be represented only as fully retarded, but not as fully advanced, that the elementary radiation processes are asymmetric. Notice that he did not maintain that the elementary processes *are* symmetric, but only that their being asymmetric *does not follow* from certain facts about the way the total field is represented mathematically.

This discussion suggests two points which we will investigate in much more detail below. First, a particular mathematical representation of the field need not be a reliable guide to the underlying physical situation, but may be adopted for reasons of mathematical convenience. Einstein's criticism of Ritz seems to have been at least partly that Ritz wanted immediately to read off the physical situation from the form of the mathematical representation. And second, we need to distinguish carefully between the *total* electromagnetic field and the *elementary processes* that give rise to the field. Thus, paralleling a distinction I drew in the last section, we have to distinguish the question of whether there is an asymmetry in possible representations of the total electromagnetic field—that is, whether the Sommerfeld radiation condition of zero incoming free fields holds, but not the condition of zero outgoing fields—from the question of whether elementary radiation processes are asymmetric (that is, the question of whether the field associated with an individual charge is asymmetric).

Distinguishing between an asymmetry of the total field and a possible asymmetry of the elementary processes contributing to the total field also allows us to make sense of Ritz and Einstein's letter (with one caveat). Einstein there wanted to contrast his own view with one that tries to reduce the arrow of thermodynamics to that of radiation. This reduction fails, one might argue, because thermodynamic or statistical reasons are needed to account for the fact that we do not observe converging spherical waves, despite the fact that they are mathematically possible. Even if elementary radiation processes are asymmetric (and result in diverging waves), spherically converging waves are physically possible. Yet in order to realize such a converging wave, a very large number of coherently oscillating elementary charged objects would be needed. But, Einstein apparently maintained, the existence of such coherent oscillators is highly improbable. Thus, the asymmetry of the elementary processes alone is not sufficient to account for the asymmetry characterizing total fields due to large numbers of accelerating charges but only in conjunction with statistical reasons. Again, this is a theme that we will explore in some detail in the next chapter. For now I just want to emphasize that the view I have in mind here does not attempt to derive an asymmetry of the fields on the elementary level from statistical considerations, but simply assumes that asymmetry as a given, and then argues that the asymmetry characterizing the total fields can be explained by appealing to the elementary asymmetry in conjunction with statistical arguments.

The only problem with attributing this view to Einstein in trying to avoid the apparent conflict between Einstein (1909b) and Ritz and Einstein (1909) is that in the latter Einstein says that the asymmetry is *"exclusively* based on reasons of probability." And admittedly I do not know how to square this exclusivity claim with Einstein's later claim that the elementary radiation processes are asymmetric.

The view I am attributing to Einstein is expressed less ambiguously, to my mind, by Popper in a series of brief letters to the Journal *Nature* (Popper 1956a, 1956b, 1957, 1958). In his first note Popper (1956a) argued that the process of waves spreading on a surface of water after a stone is dropped, exhibits an irreversibility that is distinct from the thermodynamic asymmetry. The reverse process of circularly converging waves, according to Popper, "cannot be regarded as a possible classical process." He went on to say that

> [the reverse process] would demand a vast number of distant coherent generators of waves the coordination of which, to be explicable, would have to be shown, in [a film depicting the process], as originating from the centre. This however, raises precisely the same difficulty again, if we try to reverse the amended film. (Popper 1956a, 538)

Price claims that Popper's position is similar to his own (and to the position Price attributes to Einstein), in that Popper believes the thermodynamic asymmetry and the radiation asymmetry share a common explanation (Price 1996, 52). But this strikes me as a misinterpretation of Popper's position. Popper claimed that "irreversible classical processes exist" and he drew a clear contrast between this irreversibility and that of thermodynamics: "On the other hand, in statistical mechanics all processes are in principle reversible, even if the reversion is highly improbable" (Popper 1956a, 538). Thus, for Popper there was a crucial difference between the two asymmetries: The radiation asymmetry is a *strict* asymmetry, while the thermodynamic asymmetry is not. This difference must be reflected in differences in the explanations of the asymmetries.

The asymmetry is strict even though, as Popper points out, it is not "implied by the fundamental equations" of the theory—the Maxwell equations (Popper 1956a). Popper (1957) distinguishes what he calls "theoretical reversibility" from "causal reversibility" and characterizes a radiation process as "a process that is (*a*) 'theoretically reversible', in the sense that physical theory allows us to specify conditions which would reverse the process, and at the same time (*b*) 'causally irreversible', in the sense that it is causally impossible to realize the required conditions" (Popper 1957, 1297). That is, while the Maxwell equations are time-symmetric and allow for both diverging and converging waves associated with a radiating source, converging waves violate a causal constraint; they are "causally, and therefore physically, impossible" (Popper 1958, 403). I take it that Popper's distinction, like Einstein's, is suggestive of the distinction I drew in the last section: As far as the mathematical representations of the field in terms of an initial-value or final-value problem are concerned, radiation processes are symmetric or reversible, depending on whether the fields are represented in terms of an initial- or final-value problem, the fields mathematically associated with charged particles are either retarded or advanced. And for every actual radiation process with its initial fields there is a possible process which has those initial fields as its final, outgoing fields. The Maxwell equations as dynamical equations cannot explain why the time reversals of actual radiation processes do not occur. In order to account for the asymmetry, we need to invoke an additional physical or "causal" constraint: The field physically associated with an elementary radiation process is fully retarded, and hence asymmetric.

It is worth pointing out, however, that Popper's distinction between what is theoretically possible and what is causally possible, conflicts with his own view on the notion of causation in his (1959, 1994). He maintains: "To give a *causal explanation* of an event means to deduce a statement which describes it, using as premises of the deduction one or more *universal laws*, together with certain singular statements, the *initial conditions*" (Popper 1959, 59; italics in original). And further on he says, "The 'principle of causality' is the assertion that any event whatsoever *can* be causally explained—that it *can* be deductively predicted" (Popper 1959, 61; italics in original).

Thus, Popper seems to have shared Russell's worries about a 'weighty' notion of cause, and his own account, like Russell's views, leaves no room for a distinction between what follows nomically from certain initial conditions and what is causally possible.[14] Yet there is no asymmetry of deductive prediction for wave phenomena. The Maxwell equations allow us to predict what, in light of specific initial *or final* conditions, is physically possible. And Popper tells us that making this kind of prediction is just *what we mean* by explaining an event causally. But then it becomes a mystery how radiation processes can be theoretically reversible, in that they are both describable in terms of an initial-value and a final-value problem with the help of time-symmetric laws, *and at the same time* can be causally irreversible. Against Popper's causal reductivism (and, I take it, in agreement with the view Popper appears to express in his letters to *Nature*) I will argue that the retardation condition— the claim that the field physically associated with a charge is fully retarded—plays an important explanatory role and is a causal condition in a different, 'weightier' sense that is not reducible to a claim of nomic dependency.

Popper says that the asymmetry of wave phenomena is strict. But should it not in principle be possible to set up coherent generators that produce a concentrically converging wave? I think that in order to make sense of Popper's view, we again need to distinguish carefully between the total field and the elementary processes that give rise to that field. Popper's view appears to be that the individual generators of the field are asymmetric, and this asymmetry is a strict causal asymmetry. This asymmetry can then help to explain the asymmetry of the total field. The latter asymmetry, however, is not strict as far as local field regions are concerned: If we were to set up things just right, we could produce a converging wave. But if we included the generators of the converging wave in the picture, the original asymmetry is again restored: The converging wave is shown to be the result of multiple diverging waves. Thus, in his second letter Popper says that "we are led to an infinite regress, if we do not wish to accept the coherence of the generators as an ultimate and inexplicable conspiracy of causally unrelated conspirators" (Popper 1956b, 382).

Popper's argument for why converging waves do not occur involves an appeal to initial conditions: Initial conditions necessary for the existence of converging waves virtually never obtain. But one should note that the role of initial conditions in this argument (and in the similar argument I attributed to Einstein) is rather different from that in standard arguments for the thermodynamic asymmetry. In the latter case, the asymmetry of entropy flow is accounted for by postulating what are, under some standard measure, extremely improbable initial conditions. Some have argued that this presents a problem for the standard account, since it has to leave the improbable initial conditions unexplained. But there is no similar problem for Popper's account, since most 'initial' configurations of generators will not lead to coherent converging waves. Thus, while the thermodynamic case relies on an appeal to improbable initial conditions, Popper only needs to assume that very probable initial conditions obtain—initial conditions where the individual generators are not delicately coordinated. Moreover, since the asymmetry in the case of wave phenomena is ultimately explained by an appeal to an asymmetry characterizing the individual wave generators, there need not be any asymmetry between initial and final arrangements of generators. Popper's discussion of wave generators obviously echoes Einstein's discussion of elementary processes. And Popper (1956b) explicitly refers to Einstein's discussion, saying, however, that he had not been aware of Einstein's paper when he wrote the first letter (Popper 1956a).

In support of his own reading of Popper, Price claims that Popper "noted that he was unaware of the Ritz–Einstein debate at the time he wrote his letter to Nature" (Price 1996, 270). The implication is that Popper was aligning himself with the view Einstein might have appeared to have expressed in that debate. But this strikes me as a misrepresentation of what Popper says. Clearly in reference to the generator argument, Popper says that "Einstein used a somewhat similar argument" (Popper 1956b) and then footnotes Einstein's discussion of the asymmetry of elementary processes (Einstein 1909b, 821) that I quoted above. This paper appeared after the letter concluding the Ritz–Einstein debate (Ritz and Einstein 1909), and Einstein (1909b) does not mention Ritz. There is no reference in Popper's letters to the Ritz–Einstein debate. The view with which Popper is explicitly aligning himself is the view that "the reverse process [of a spherically diverging wave] does not exist as elementary process," and not the view that "the irreversibility is exclusively based on reasons of probability."

Price suggests that Popper might have taken his generator argument as an argument for the asymmetry of the elementary processes, instead of assuming this asymmetry as a premise in order to account for the asymmetry of the total fields, as I have suggested. As far as I can see, it is not unequivocally clear from the text whether Popper meant to use the regress argument to establish the causal asymmetry or whether he used the claim that there is a causal asymmetry as a premise in his argument. My own reading is partly motivated by a principle of charity, because Price's criticism of any such attempt to argue for the causal asymmetry is cogent.[15] To claim that the existence of coherently oscillating wave generators producing a converging wave would constitute an inexplicable conspiracy is to presuppose that the fields associated with generators are fully retarded. For to assume that only retarded fields, and not advanced fields (if they were to exist), could be the source of coherent motion is to be guilty of a temporal "double standard fallacy" (Price 1996, 55). Popper himself may well have been guilty of a temporal double standard. Yet my interpretation proposes a way to avoid the fallacy: by assuming as a premise that the elementary wave processes are time-asymmetric.

In this section I have argued that while Einstein and Popper explored the connections between the wave asymmetry and that of thermodynamics, neither of them seems to have believed that the former asymmetry can be explained by appeal to the latter alone, and both appeal to a nonthermodynamic asymmetry of individual wave generators. Before I discuss and criticize thermodynamic and cosmological accounts of the radiation asymmetry in the next chapter, I want to end this chapter by briefly discussing two putative explanations of the asymmetry in which thermodynamic considerations play no role at all.

4. Rohrlich on the Asymmetry of an Equation of Motion

Rohrlich has argued that the problem of the temporal asymmetry exhibited by electromagnetic phenomena can be solved if one considers the differential-difference equation of motion for a relativistically rigid charged shell instead (Rohrlich 1998, 1999, 2000). As we have seen in chapter 3, if one ignores non-linear terms, the equation governing a small charged shell is

$$m\dot{\mathbf{v}} = \mathbf{F}_{ext} + \frac{e^2}{3c\,a^2}\left[\mathbf{v}\left(t - \frac{2a}{c}\right) - \mathbf{v}(t)\right], \tag{3.8}$$

where a is the radius of the charge. Like Dirac's theory (and unlike the Lorentz equation of motion discussed in chapter 2) this equation takes into account self-interaction effects.

As Rohrlich points out, this equation is not time-reversal-invariant. Rohrlich takes equation (3.8) to be the fundamental particle equation of motion of classical electrodynamics, and thinks that this solves the problem of the temporal asymmetry of electromagnetic phenomena. For while the Maxwell equations are time-symmetric, the particle equation of motion, which includes self-interactions, is not: "The *problem of the arrow of time in classical dynamics* is therefore solved: the fundamental equations of classical physics are *not* time reversal invariant" (Rohrlich 2000, 11).

Rohrlich's account, however, simply sidesteps, rather than answers, the traditional puzzle of the arrow of radiation. The traditional problem concerns an asymmetry of radiative phenomena that arises for a given configuration of sources. Thus, the puzzle concerns a putative asymmetry of radiation *fields*, and it arises independently of the question of what the proper equation of motion governing the sources is. In a retarded or advanced field problem, the trajectories of the sources are simply assumed to be given. How a time-asymmetric particle equation of motion such as (3.8) can be derived from the time-symmetric Maxwell equations and the principle of energy–momentum conservation is a separate question. Rohrlich's account does provide an answer to this question. Rohrlich explains:

> [the origin of this noninvariance] is physically intuitive because self-interaction involves the interaction of one element of charge on the particle with another such element. That interaction takes place by the first element emitting an electromagnetic field, propagating along the *future light cone*, and then interacting with another element of charge. The future light cone (rather than the past light cone) was selected using the *retarded* fields (rather than the advanced fields). An

asymmetry in time was thus introduced according to the *causal* nature of the process. (Rohrlich 2000, 9 italics in original)

Thus, equation (3.8) is asymmetric, because in addition to the Maxwell equations one assumes in the derivation that the interactions between different parts of the charge are fully retarded. The asymmetry of the equation of motion is explained by an appeal to an asymmetry of the fields.

Rohrlich's account does not, however, answer the problem of the asymmetry of radiation. For, first, the account does not even mention the asymmetry between initial and final conditions embodied in the condition RADASYM. Rohrlich explains a putative asymmetry in the *equation of motion*, but not one in the *fields*. Second, the account appeals to the field physically associated with a charge and assumes that this field is fully retarded. But according to one version of the puzzle, it is precisely the latter asymmetry that is in need of an explanation: Given that the Maxwell equations are time-symmetric, why can the total field usually be represented as fully retarded? And is it legitimate as a solution to the puzzle to appeal to the notion of a retarded field physically or causally associated with an element of charge? According to an influential view, as we shall see, the answer to this last question is 'no.' Rohrlich, however, says nothing to address the first question, and in his account of why the particle equation of motion is asymmetric, simply assumes without argument that the answer to the second question is 'yes.' While I agree with Rohrlich that the best account of the asymmetry has to invoke a fundamental causal constraint, in light of the seemingly powerful arguments against such an account, more has to be said to show that such an account can indeed offer a legitimate solution to the puzzle.

5. Anderson's Finite Energy Requirement

James Anderson (1992) has argued that there is no arrow of radiation, in that neither retarded nor advanced representations of the field are privileged. We pick the retarded representation when a problem is set up in terms of initial conditions, and pick the advanced representation when a problem is set up in terms of final conditions. So far, of course, everyone would agree. The question, however, is why the retarded representation generally satisfies the Sommerfeld radiation condition, according to which the field is *purely* retarded, with no contribution from a free field satisfying the homogeneous wave equation. Anderson addresses this question by arguing that "the [purely] retarded potentials arise naturally when one solves the wave equation as an initial [as opposed to final] value problem and imposes the physical requirement that the initial energy in the wave field is finite" (Anderson 1992, 466). For, as he shows, if the total energy of the free field F_{in} over *all* of space is finite at the initial time t_0, then $\lim_{t \to \infty} F_{in} = 0$. An initially finite field will eventually spread through infinite space so that in the limit the field will go to zero at each point in space. Thus, in the remote future, Anderson argues, the total field at each point will be given by the retarded field alone, with no contribution from the free field.

Can this argument solve the puzzle of the arrow of radiation? Recall that the puzzle is why the Sommerfeld radiation condition of zero incoming fields holds approximately *but* the condition of zero outgoing fields does *not*. Yet at most Anderson can show why incoming fields are often approximately zero, since his argument is time-symmetric and does not single out the retarded field representation over a representation in terms of advanced fields. As Anderson himself points out, his argument also shows that in the remote past, any finite free outgoing field F_{out} was zero. That is, independently of whether we set up an initial-value or final-value problem—that is, no matter which field representation we choose—and no matter what the initial or final fields are (as long as they are finite), sufficiently far from the hyperplane on which the initial or final values are specified, the field will be entirely due to the sources and will be fully retarded or fully advanced, respectively.

Anderson argues that at many (or even at most) times the condition of zero incoming fields holds. Even if at some time t_0 the incoming field is nonzero, the source-free Maxwell equations predict that the free field will eventually tend to zero at later times. And, similarly, this free field will have been zero in the remote past. Notice that it follows from the symmetric character of the argument that a field which at one time is completely dispersed with a field strength that is nowhere finite, can at later times be nonzero. We can see this if we focus on the past counterpart to the argument. The argument shows that a free field which in the infinite past was completely dispersed (with zero field strength everywhere) can at a finite time have nonzero field strengths within a certain region. Thus, even if a finite free field dispersed very quickly, such a field could at some finite time in the future result in regions with finite field values.

How, in light of Anderson's symmetric argument, could we account for the perceived asymmetry (or at least for what we perceive to be an asymmetry) of radiation phenomena? In presenting the puzzle of the arrow of radiation, Zeh asks, "Why are initial conditions more useful than final conditions?" So far Anderson's argument provides no reason why initial conditions should be more useful. Both retarded and advanced representations require the *entire* trajectories of the relevant charges as inputs, in addition to the initial or final fields, respectively. And if, as Anderson seems to show, it is as easy to choose final times during which the outgoing fields are zero as it is to choose initial times with zero incoming fields, there seems to be no pragmatic advantage for retarded representations.

One source of an asymmetry that may be able to explain our preference for retarded fields might be a difference in the amount of time it takes for free fields to become negligible in the two temporal directions. That is, it might be the case that fields tend to zero very quickly toward the future, while nonzero fields would have evolved from fields that had been zero only in the remote past. Thus, it might in fact be much easier to represent a phenomenon in terms of a purely retarded field than in terms of a purely advanced field. Anderson himself suggests that there might be such an asymmetry:

> Almost all initial conditions lead very quickly to irreversible motion [where the irreversibility is supposed to be "a consequence of the finite energy requirement"] as everyone who has ever thrown a stone in a pond can attest. It is however extremely hard to set them up so that an incoming wave results for any length of time. (Anderson 1992, 467)

As far as I understand Anderson here, he seems to claim that purely retarded waves result almost instantly under almost all initial conditions, while purely advanced waves are extremely rare. Thus, he seems to concede that there is an asymmetry between retarded and advanced representations after all. But this asymmetry is left unexplained by his argument, which treats initial-value and final-value problems completely on a par.

Any account of a difference between how fast free fields tend toward zero in the two temporal directions cannot, of course, be presented in terms of the source-free Maxwell equations alone, and presumably would need to appeal to the properties of some material medium to which the fields are coupled (or, within the context of the general theory of relativity, to the geometric properties of space-time). One very plausible suggestion is that the difference is due to the presence of absorbing media. And in fact appeals to various types of absorbers feature prominently in a number of accounts of the asymmetry of radiation, which will be the focus of the next chapter.

6. Conclusion

Radiation phenomena and wave phenomena more generally seem to exhibit a temporal asymmetry. In this chapter I proposed the following formulation of this asymmetry:

> (RADASYM) There are many situations in which the total field can be represented as being approximately equal to the sum of the retarded fields associated with a small number of charges (but not as the sum of the advanced fields associated with these charges), and there are almost no situations in which the total field can be represented as being approximately equal to the sum of the advanced fields associated with a small number of charges.

Importantly, this asymmetry is one between prevailing initial and final conditions, and not that of an asymmetric dynamical constraint.

I then examined discussions of the 'puzzle of the arrow of radiation' by Einstein and Popper, partly with the aim of correcting certain misinterpretations of their views in the literature and partly of introducing several themes to which I want return in chapter 7 when I propose my own answer to the puzzle. Both Einstein and Popper appeal to what Einstein calls "elementary radiation processes" and to statistical considerations. As I will argue, the most successful solution to the puzzle does in fact invoke both notions.

I ended the chapter with a brief discussion of two purported solutions to the puzzle that lie somewhat outside the main literature on the subject, which seems to have focused on the role of absorbers in accounting for the asymmetry: Rohrlich's discussion of an asymmetric particle equation of motion and Anderson's appeal to a finite energy requirement. Rohrlich's account, to my mind, correctly appeals to the claim that individual charges are associated with diverging radiation, but does not discuss how this might affect an asymmetry between initial and final conditions and does not justify his appeal to a fundamental causal asymmetry. Anderson's arguments are completely time-symmetric, and therefore cannot account for any time asymmetry.

6

Absorber and Entropy Theories of Radiation

1. Introduction

Absorber theories of radiation take as their starting point John Wheeler and Richard Feynman's action-at-a-distance theory with symmetric interactions between particles in a region surrounded by an infinite absorber (Wheeler and Feynman 1945). Common to all such theories is the idea that the asymmetry of radiation is due to certain nonelectromagnetic properties of an absorber with which radiating sources interact and, hence, that the arrow of radiation is reducible to some other physical asymmetry. Wheeler and Feynman themselves argue that the asymmetry is due to statistical reasons, while Paul Davies (1974) and H. Dieter Zeh (2001), whose account I will discuss in chapter 7, attempt to derive the asymmetry from an asymmetry of phenomenological thermodynamics. J. E. Hogarth (1962) and F. Hoyle and J. V. Narlikar (1995) claim that the asymmetry can be derived from the cosmological asymmetry of an expanding universe. Huw Price argues that the "mathematical core" of Wheeler and Feynman's theory can be used to derive the asymmetry from asymmetric cosmological initial conditions and that the asymmetry thus has the same origin as the thermodynamic asymmetry of entropy increase (Price 1996; see also Price 1991a, 1991b, 1994).

In their paper, Wheeler and Feynman present two distinct conceptions of an ideal absorber. The first is based on a general, formal definition of an absorber, stipulating that fields on the outside of an absorber enclosing a radiating source are zero. The second conception relies on a physical model of an absorbing medium as dilute plasma in which fields are exponentially damped in the forward time direction. Both conceptions have been adopted in other absorber theories. Since the two conceptions play distinct roles in various arguments for how the radiation asymmetry can arise in a symmetric theory, I will discuss them separately.

I will begin by discussing Wheeler and Feynman's argument, based on the formal conception of an absorber, for why their symmetric action-at-a-distance theory is equivalent to a retarded particle–field theory. I will argue that this attempt to account for the radiation asymmetry fails. The formal conception also plays an important role in attempts to reduce the radiation asymmetry to a cosmological asymmetry. Hogarth, and Hoyle and Narlikar, argue that a fully retarded representation of the total fields is picked out uniquely if the absorber is asymmetric, in that only the future, but not the past, is ideally absorbing. This asymmetry, they argue, is a feature of certain cosmological models that postulate an expanding universe. Yet, as I will show, these arguments are unsuccessful as well.

I will then turn to a discussion of Wheeler and Feynman's specific model of an absorber as dilute plasma, which also plays a prominent role in the theories of Hogarth and of Hoyle and Narlikar, as well as in Davies's attempts to reduce the radiation asymmetry to that of thermodynamics. I will argue that despite its widespread use in symmetric action-at-a-distance theories, the model is in fact incompatible with the assumption of a time-symmetric half-retarded, half-advanced field associated with charged particles.

Then I will examine Huw Price's solution to the puzzle of the arrow of radiation, which is based on an attempt to reinterpret the mathematical core of the Wheeler–Feynman theory. For Price the apparent asymmetry of radiation is purely a macroscopic phenomenon that is due to cosmological initial conditions and has the same origin as the thermodynamic asymmetry. However, as I will show, Price's arguments for the microsymmetry and macroasymmetry of radiation fail.

Thus, the overarching thesis of this chapter is a negative one: Ultimately, attempts to reduce the radiation asymmetry to another temporal asymmetry within the context of an absorber theory are unsuccessful.

2. Wheeler and Feynman's Formal Absorber Theory

Wheeler and Feynman present their absorber theory as an alternative formalism for deriving the Lorentz–Dirac equation of motion for a charged particle (Wheeler and Feynman 1945). I discussed the difficulties of arriving at a conceptually unproblematic equation that is consistent with the Maxwell equations and the principle of energy conservation in chapters 2 and 3. I said there that the most promising candidate for a consistent point particle equation of motion is the Lorentz–Dirac equation (see Dirac 1938). Unlike the Lorentz force law, Dirac's equation takes into account the radiative reaction of the charge—that is, the effect of the charge's own field on the motion of the charge. Wheeler and Feynman endorse the Lorentz–Dirac equation and offer their theory as an alternative framework for deriving the equation that is intended to avoid some of the conceptual problems associated with Dirac's theory.

Recall that the difficulty in trying to include the radiative reaction is that the field of a charge is infinite at the location of the charge. In order to overcome this difficulty, Dirac assumes that all actual fields associated with charges are fully retarded, but then rewrites the retarded field of a charge i as

$$F^i_{ret} = 1/2(F^i_{ret} + F^i_{adv}) + 1/2(F^i_{ret} - F^i_{adv}). \tag{6.1}$$

The term in the first set of parentheses on the right-hand side is the problematic infinite term. Since this term formally acts like a mass term in the equation of motion, Dirac proposes that one should understand it as an electromagnetic contribution to the total empirically observable mass, which nevertheless is finite. The second term in (6.1), which is finite, represents the radiative reaction. This term can be calculated explicitly and, as we have seen, depends not only on the acceleration of the charge but also on the derivative of the acceleration. The total field force acting on a charge is given by the sum of the radiative reaction, the retarded fields of all other charges, and any free incoming fields:

$$\sum_{k \neq i} F_{ret}^{k} + F_{in} + \frac{1}{2}(F_{ret}^{i} - F_{adv}^{i}). \tag{6.2}$$

The explicit form of the Lorentz–Dirac equation (4.2) need not concern us here.

The first problem with Dirac's derivation, according to Wheeler and Feynman, concerns temporal asymmetries. They take it to be unsatisfactory that Dirac assumes that the field associated with a charge (what Dirac calls the "actual field") is fully retarded, positing a temporally asymmetric interaction between charges and fields even though the Maxwell equations are time-symmetric. By contrast, the Wheeler–Feynman theory assumes that the interaction is symmetric in time, half-retarded and half-advanced. Thus, they assume that the notion of the field physically associated with a charged particle is a legitimate notion, even though they disagree with the standard assumption that the field associated with a charge is fully retarded.

The second problem of Dirac's theory concerns the treatment of the infinite self-field of a charge. Wheeler and Feynman's theory avoids any problems associated with the pathological infinities and Dirac's proposed 'cure' of renormalizing the mass by assuming that the electromagnetic force on a charge is due only to the fields of all other charges. In their theory there is no self-interaction, and hence no infinities need to be 'swept under the rug.' Wheeler and Feynman get rid of self-interactions by denying that the electromagnetic field is a real entity, ontologically on a par with material objects. On their view, the field is merely a mathematical auxiliary device designed to track the forces that charged particles exert on one another at a distance. Now, a problem with retarded action-at-a-distance theories of electrodynamics had been that they violate Newton's second law: If the interaction between charged particles is retarded (and hence asymmetric), then the action of one charge on another is not in general balanced by an equal and opposite reaction. Yet in a theory with time-symmetric interactions between particles, this problem does not arise.

The task, then, for Wheeler and Feynman is to show that a symmetric theory without self-interaction can result in the same particle equation of motion as Dirac's asymmetric theory with self-interaction. In Dirac's theory the radiation reaction term is a result of the interaction of a charge with its own field. Wheeler and Feynman show that the same term can be derived in their theory, if one postulates that the charge is surrounded by an infinite absorber.

In their paper they offer four different derivations of the equivalence of their theory with Dirac's classical theory of the electron, all of which rely crucially on assumptions about infinite absorbers. The first three derivations introduce specific physical modeling assumptions regarding the absorber medium in order to derive explicitly the contribution of the absorber particles to the total field. The fourth and most general derivation, on which I want to focus first, assumes only a general, formal definition of an absorbing medium.

If a charge i is surrounded by an infinite absorber, then all fields vanish outside the absorber. Thus, Wheeler and Feynman formally define an absorber by the relation

$$\sum_k F^k_{ret} + \sum_k F^k_{adv} = 0 \quad \text{(outside the absorber)}, \quad (6.3)$$

where the sum ranges over all charges k, including both the charge i and the charges constituting the absorber. Defined in this way, the absorber is time-symmetric. That is, the absorber ensures that there are no nonzero fields either in the distant future or in the remote past of the charge i. As far as the advanced field associated with the charge is concerned, this means that the 'absorber' acts as what one would normally take to be an *anti-absorber*, if looked at in the normal time sense: There is no nonzero field in the past of the absorber, even though there is a nonzero advanced field (associated with the charge i) in the interior of the absorbing region. Looked at in the forward time sense, this field seems to be arising within the absorber.

Since the retarded fields represent an outgoing wave and the advanced fields represent an incoming wave, and complete destructive interference between two such waves is impossible, the two sums have to vanish individually:

$$\sum_k F^k_{ret} = 0 \quad \text{and} \quad \sum_k F^k_{adv} = 0 \quad \text{(outside the absorber)}, \quad (6.4)$$

which implies that the difference of the two fields vanishes as well:

$$\sum_k (F^k_{ret} - F^k_{adv}) = 0 \quad \text{(outside the absorber).} \quad (6.5)$$

As Dirac has shown, the difference between the retarded field and the advanced field is a source-free field. But if a source-free field vanishes somewhere, it has to vanish everywhere and not just outside the absorber. Thus,

$$\sum_k (F^k_{ret} - F^k_{adv}) = 0 \quad \text{(everywhere).} \quad (6.6)$$

Equation (6.6) represents the central property of the ideal absorber that enables Wheeler and Feynman to derive the equivalence of their symmetric theory with Dirac's theory involving fully retarded fields.

The field exerting a force on a point charge i surrounded by an infinite absorber, which by assumption is given by the sum of the half-retarded, half-advanced fields of the absorber particles, can be rewritten as follows:

$$\frac{1}{2}\sum_{k\neq i}(F^k_{ret} + F^k_{adv}) = \sum_{k\neq i} F^k_{ret} + \frac{1}{2}(F^i_{ret} - F^i_{adv}) - \frac{1}{2}\sum_{\text{all } k}(F^k_{ret} - F^k_{adv}). \quad (6.7)$$

Here the sum on the left-hand side and the first sum on the right-hand side range over only the absorber particles; that is, over all particles except the charge i. According to (6.6), the second sum on the right-hand side vanishes, so that the field force acting on i is

$$\sum_{k \neq i} F^k_{ret} + \frac{1}{2}(F^i_{ret} - F^i_{adv}). \tag{6.8}$$

The first term represents the familiar external retarded fields due to all other charges, and the second term is the radiative reaction term which Dirac had calculated explicitly. Equation (6.8) is equivalent to (6.2), the equation derived by Dirac, if one assumes, as do Wheeler and Feynman, that there are no free incoming fields. Thus, Wheeler and Feynman show how one can derive an asymmetric, retarded equation of motion from time-symmetric laws. Moreover, it follows from (6.8) that if we add the symmetric field of the charge i, the external field on a small test charge j in the vicinity of i is equal to the sum of the fully retarded fields of the absorber particles and of i:

$$\sum_{k \neq i} F^k_{ret} + \frac{1}{2}(F^i_{ret} - F^i_{adv}) + \frac{1}{2}(F^i_{ret} + F^i_{adv}) = \sum_{all\ k} F^k_{ret}. \tag{6.9}$$

One might think that Wheeler and Feynman have succeeded in showing, starting from time-symmetric laws, how it is that the field associated with a charge appears to be fully retarded. But as it stands, the argument does not yet show that radiation is asymmetric. For the argument cannot establish why the radiation associated with a charge appears to be fully retarded *rather than advanced*, since from Wheeler and Feynman's assumptions of a time-symmetric interaction and an infinite absorber one can equally derive that the field acting on the charge i can be given the advanced representation

$$\sum_{k \neq i} F^k_{adv} - \frac{1}{2}(F^i_{ret} - F^i_{adv}) \tag{6.10}$$

and, similarly, that the field acting on a small test charge j is the fully advanced field $\sum_k F^k_{adv}$. Thus, Wheeler and Feynman have not yet located an asymmetry that could count as a solution to the puzzle of the arrow of radiation.

They are, of course, aware of this, and argue that the asymmetry between the two representations is due to statistical considerations: "We have to conclude with Einstein that the irreversibility of the emission process is a phenomenon of statistical mechanics connected with the asymmetry of initial conditions with respect to time" (Wheeler and Feynman 1945, 170).[1] For, Wheeler and Feynman argue, before the source i turns on, the absorber particles will be in random motion or at rest, so that the retarded absorber field in (6.8) and (6.9) can be taken to be equal to zero. By contrast, "the sum of the advanced field of the absorber particles [in (6.10)] is not at all negligible for they are put in motion by the source at just the right time to contribute to [that sum]" (ibid.). Moreover, since (6.8) and (6.10) are equivalent, it follows from the assumption that the retarded field of the absorber is zero, that the advanced absorber field is equal to twice the radiation reaction force, canceling the negative reaction force in (6.10). Thus, despite appearances, the total force on the charge i in (6.10) is equal to the positive radiation reaction force as well.

The argument would also show why the field associated with the charge i seems to be fully retarded. The field acting on a test charge j would in the retarded representation consist only of the retarded field of the source i, since the retarded field of the absorber is zero, while in the advanced field representation one would have to take into account the advanced field of all the absorber particles as well, in addition to the advanced field of i (where these fields would of course conspire to give the fully retarded field associated with i).

But Wheeler and Feynman's argument is, as Price has shown, guilty of a "temporal double standard" fallacy (Price 1996, 68).[2] Frequently, Price maintains, arguments for temporal asymmetries from seemingly time-symmetric premises smuggle in a bias for one temporal direction. Wheeler and Feynman's statistical argument for the asymmetry is a paradigm example of this mistake. They argue that the sum of the advanced fields of the absorber particles is not negligible, since the retarded field of the source i sets the absorber particles into coherent motion. Without the radiating source it would be extremely improbable for the absorber particles to oscillate coherently. In the presence of a source, however, we would expect correlated motions among the absorber particles due to the retarded field of the source. So far so good. Yet if Wheeler and Feynman's statistical considerations apply in one temporal direction, they ought to apply in the other direction as well. If the radiation due to i is in fact half retarded and half advanced, as Wheeler and Feynman assume, then the advanced component of that field should result in correlated motions of the absorber particles equivalent to those due to the retarded component. Thus, there should be a nonzero retarded 'response' wave as well. If the field associated with the source i is truly symmetric and if the retarded field of the source leads to correlated motions in the absorber, then so should the advanced field of the source. Both components of the field associated with the source ought to have equivalent effects on the absorber. But then the retarded absorber field in (6.8) would be nonzero in general, and Wheeler and Feynman's argument for why a negative radiation reaction force and fully advanced fields cannot be observed does not go through. Thus, Wheeler and Feynman's appeal to statistical considerations cannot show how it is that all radiation produced by sources appears to be fully retarded rather than fully advanced.

3. Asymmetric Absorbers and Cosmology

Hogarth (1962) argues that the radiative asymmetry is a result of a cosmological asymmetry between past and future absorbers. Similar arguments are advanced by Hoyle and Narlikar (1995). According to Hogarth, the indeterminacy between retarded and advanced representations that plagues Wheeler and Feynman's theory arises only if both the future and the past are ideally absorbing. If only the future absorber is ideal, while the past absorber is not, then the retarded representation is picked out uniquely. By contrast, if only the past absorber is ideal, then the advanced representation is the correct one. This argument invokes Wheeler and Feynman's formal conception of an absorber. In the second step of the argument Hogarth (and Hoyle and Narlikar) appeal to Wheeler and Feynman's physical model of the absorber to argue that whether the past or the future is absorbing depends on the specific cosmological model and is independent of any thermodynamical considerations. Wheeler and Feynman had set up their theory within the static universe of special relativity. If, by contrast, we assume that the universe is expanding, the symmetry between past and future absorbers can be broken. A difference in the efficiency of the absorbers can arise, since in an expanding universe, light traveling toward the future is redshifted, while light traveling toward the past (as advanced radiation) is blueshifted. In addition, the density of matter is not constant in an expanding universe.

There has been some argument over whether Hogarth's account is in fact completely independent of thermodynamic considerations.[3] There is, however, a more fundamental problem with these proposals to reduce the electromagnetic asymmetry to that of cosmology: The arguments within the formal absorber theory for the claim that an asymmetry between past and future absorbers uniquely picks out either a retarded or an advanced field representation are flawed.

Crucial to the first part of Hogarth's argument is the claim that if only one absorber is ideal, then the other, nonideal absorber effectively drops out of the picture and the field acting on the source will be completely independent of that second absorber. The argument is this. Hogarth begins by relating the fields entering an absorbing region—what he calls the *stimulus*—to the *response* of the absorber. If we define

$$F = 1/2(F_{ret}^i - F_{adv}^i), \tag{6.11}$$

where F_{ret}^i and F_{adv}^i are respectively the retarded and advanced fields associated with a charge i in an absorber cavity, then "the stimulus of the future absorber is $F + \Pi$, where Π is the response of the past absorber, and the stimulus of the past absorber is $-F + \Phi$, where Φ is the response of the future absorber" (Hogarth 1962, 371). This is so because the field entering the future absorber is the sum of the field associated with the source and the field propagating from the past absorber into the future; and similarly for the past absorber. Of course, as Hogarth points out, in the region of the future absorber, $F = 1/2F_{ret}^i$, while in the region of the past absorber, $F = -1/2F_{adv}^i$. Thus, the total field acting on the future absorber will be given by the sum of the retarded fields of the past absorber and the source i, which will result in an advanced response wave Φ; and the field acting on the past absorber will be the sum of the advanced fields of the future absorber and the source, which in turn will result in a retarded response wave Π. Hogarth relates the stimulus and response fields to one another by writing

$$\Phi = f(F + \Pi) \tag{6.12}$$

and

$$\Pi = p(-F + \Phi), \tag{6.13}$$

where f and p are operators which map the stimulus fields onto the response fields. Hogarth then claims that for an ideal absorber the relevant operator is equal to 1, while in the case of a nonideal absorber the operator will be less than 1. Thus, if the future absorber is ideal, but the past absorber is not, $f = 1$ and $p \neq 1$, which implies that $\Phi = F$ and $\Pi = 0$. Since the radiative reaction—which in the action-at-a-distance theory is equal to the response of the absorbers on the source—is $F_{rad} = \Phi + \Pi$, it follows that

$$F_{rad} = F = 1/2(F^i_{ret} - F^i_{adv}). \tag{6.14}$$

If we compare this result with (6.8) and (6.10) above, we see that the radiation reaction is that of the retarded field representation. Thus, Hogarth concludes, if the future absorber is ideal while the past absorber is not, the retarded field representation is uniquely singled out.

Crucial to the argument is the claim that the operators f and p are equal to unity for ideal absorbers and not equal to unity otherwise. Unfortunately, Hogarth offers no justification for this assumption, but I think one might try to justify it as follows. An ideal future absorber is a medium for which there are no nonzero fields at its future boundary. That is, the fields satisfy the condition $\sum_k F^k_{ret} = 0$ in the absorber's future (which is half of (6.4)). Hence, if we solve the wave equation as a final-value problem for the fields in a region within the absorber and in its immediate past, and take as the final-value surface a hyperplane in the absorber's future, the free outgoing field will be zero on the boundary. The total field will be fully advanced and will be equal to twice Hogarth's response field Φ.[4] Alternatively, we could represent the same field in terms of an initial-value problem. The fields mathematically associated with charged particles will then be retarded fields, to which any free incoming field would have to be added. If we choose as initial-value surface a hyperplane in the past of the past absorber, the field will be equal to the sum of the fully retarded field of the source, the retarded field of the past absorber, and any free field coming in from past infinity. That is, if we assume that there are no free incoming fields—that is, if we assume that the past absorber is ideal as well and satisfies $\sum_k F^k_{adv} = 0$ in the absorber's past—the field is equal to $F^i_{ret} + 2\Pi$. Hence,

$$2\Phi = F^i_{ret} + 2\Pi. \tag{6.15}$$

For ideal absorbers the operator f, thus, is unity.

By contrast, if the future absorber is nonideal, the fields entering the absorbing region will not be damped out completely. If we represent the field in terms of a final-value problem, there will be an additional free field contribution to the field aside from the advanced field associated with the absorber, and therefore 2Φ will be less than the sum $F^i_{ret} + 2\Pi$. In other words, f will be less than 1.

Thus, if both absorbers are ideal, f is equal to 1, and if the future absorber is nonideal, while the past absorber is ideal, f will be less than 1. This is not, however, what Hogarth needs. Hogarth's argument relies on the claim that if the future absorber is ideal and the past absorber is nonideal, then $f = 1$, and this claim is false. The total electromagnetic field can mathematically be equivalently represented as a sum of retarded fields and free incoming fields or as a sum of advanced fields and free outgoing fields. That is, the following equality holds for the total field in the interior of the absorber cavity in the presence of a charge i:

$$\sum_{\pi} F^{\pi}_{ret} + F^{i}_{ret} + F_{in} = \sum_{\phi} F^{\phi}_{adv} + F^{i}_{adv} + F_{out}. \tag{6.16}$$

Here the index ϕ ranges over all those charges in the past absorber that lie in the future of the initial-value surface and the index π ranges over all those charges in the future absorber that lie in the past of the final-value surface. If we choose initial-value and final-value surfaces outside the absorber, then the sums range over all absorber particles, and the free incoming and outgoing fields are nonzero only if the absorber is nonideal.

It is easy to specialize to the case of an ideal future absorber and nonideal past absorber. In that case,

$$\sum_{\pi} F^{\pi}_{ret} + F^{i}_{ret} - F^{i}_{adv} + F_{in} = \sum_{\phi} F^{\phi}_{adv}, \tag{6.17}$$

which in Hogarth's notation can be written as

$$\Phi = F + \Pi + 1/2 F_{in}. \tag{6.18}$$

Comparison with (6.12) and (6.13) shows that both f and p are different from 1. The case of an ideal past absorber and a nonideal future absorber is similar. Thus, contrary to what Hogarth claims, f and p are different from 1 as long as at least one absorber is nonideal. But equation (6.18) does not allow us to determine an expression for the radiation reaction $\Phi + \Pi$ in terms of the field F associated with i.

It is worth pointing out that it would have been quite surprising, in light of Wheeler and Feynman's original derivation, if Hogarth's argument had been sound. For that derivation, we may recall, crucially relied on (6.5), according to which the difference between the total retarded fields and advanced fields vanishes outside the absorber. Yet it follows from (6.17) that in the case of a nonideal absorber this difference is nonzero.

Hogarth's mistake is that he does not allow for the presence of free fields in the case of nonideal absorbers. His argument would go through only if we assumed that even when the past absorber is nonideal, there are no free incoming fields, and this assumption is unwarranted. Note that one cannot argue that free fields in (6.16) are excluded by the fact that the Wheeler–Feynman theory is an action-at-a-distance theory. Clearly, free fields in the relevant sense here are compatible with the theory. In an action-at-a-distance theory there are no *physically* free fields—that is, no fields that are not physically associated with certain charges. But this does not conflict with the claim that certain *mathematical* representations of the field will involve source-free fields. The free incoming fields that are part of the retarded representation of the fields associated with a nonideal past absorber are physically associated with the charges in their future. The sense in which fields are present in the charge-free regions in the past of the absorber is the same sense in which we can in general speak of the presence of fields in an action-at-a-distance theory: A small test charge that was placed in the past of the past absorber would experience an electromagnetic force.

No matter what our view on the physical contribution of charged particles to the field is, electromagnetic fields have to satisfy the Maxwell equations. And from these we know that if there is a nonzero electromagnetic field acting on a test charge in the past of the past absorber, then this field will show up as a free incoming field in a retarded representation of the field. Thus, in (6.12), if the past absorber is nonideal, then f will not be equal to 1, even if the future absorber is ideal.

One might object to my argument by maintaining that the past absorber could still be an ideal absorber for *incoming* fields, even if it is does not damp out all advanced fields in the past time direction. The past absorber would be ideal if there were no advanced radiation in its past due to sources in its future. Its being *nonideal in this sense*—that is, its being transparent to advanced radiation originating in its future—is compatible with its being completely opaque to retarded radiation originating in its past, or at least so one might claim. But then the free incoming field in the absorber's past does not show up in the future of the absorber, and it seems to follow from (6.12) that f is equal to 1 after all. The problem, however, then is with (6.13). The only motivation for thinking that p ought to be less than 1—that is, the only motivation for thinking that the retarded response field of the past absorber is not equal to the sum of the advanced fields in its future—is to assume that there is an additional free incoming field in the case of nonideal absorbers. If this free field is damped out within the absorber, then the field in the immediate future of the past absorber can equivalently be represented as

$$\sum_\pi F^\pi_{ret} + F^i_{ret} = \sum_\phi F^\phi_{adv} + F^i_{adv}, \qquad (6.19)$$

from which it follows that both f and p are equal to 1.

Hoyle and Narlikar (1995) present what they claim amounts to only a slightly different version of Hogarth's argument. Like Hogarth, they argue that if the future absorber is ideal but the past absorber is not, then the retarded field representation is singled out unambiguously. Yet contrary to Hogarth's argument, Hoyle and Narlikar's relies on Wheeler and Feynman's specific absorber model. I will discuss this model in detail in the next section, but in order to be able to discuss Hoyle and Narlikar's argument here, I need to mention the central result that Wheeler and Feynman purport to establish: that the advanced response wave due to the future absorber alone is just Dirac's radiation reaction term. This means that the total field in the neighborhood of the charge, ignoring the field of the past absorber, is equal to the fully retarded field of the source. Hoyle and Narlikar generalize this result as follows. They assume, first, that the total field in the vicinity of the charge is equal to some linear combination of the charge's retarded and advanced fields:

$$F = AF_{ret} + BF_{adv}. \qquad (6.20)$$

Next they argue that since a fully retarded field entering an ideal absorber results in the Dirac radiation reaction term, a field of strength AF_{ret} entering a less than ideal absorber should result in a response field AfF_{ret}, where f is less than 1; and similarly for the reaction due to the past absorber. Taking into account that the past absorber leads to a reaction term with the opposite sign and adding the field associated with the source, the total field can thus also be written as

$$F = 1/2(F_{ret} + F_{adv}) + 1/2(Af - Bp)(F_{ret} - F_{adv}). \tag{6.21}$$

Equating the coefficients of F_{ret} and F_{adv} separately in (6.20) and (6.21) leads to the following two equations for A, B, f, and p:

$$2A = 1 + Af - Bp \text{ and } 2B = 1 + Bp - Af. \tag{6.22}$$

If both absorbers are ideal—that is, if both f and p are equal to 1, then A and B are left undetermined (since the resulting equations are linearly dependent). If, however, only the future absorber is ideal and the past absorber is not—that is, if $f = 1$ and $p \neq 1$—then it follows that $A = 1$ and $B = 0$. And similarly for an ideal past absorber with nonideal future absorber. Thus, Hoyle and Narlikar conclude, an ideal future absorber together with a nonideal past absorber singles out the retarded representation uniquely.

However, the argument contains an error similar to that committed by Hogarth. For if one of the absorbers is not ideal, then there are nonzero fields at the outside of that absorber that have an effect on the total fields in the interior of the absorber close to the source. For example, let us assume that the future absorber is not ideal. Then there is a nonzero field in the absorber's future. Calculating the field near the source in terms of a final-value problem, and again assuming an initial-value surface in the future of the absorber, requires taking into account a nonzero outgoing field F_{out}. Thus, even if the absorber particles on their own contribute a field proportional to the radiation reaction, as Hoyle and Narlikar claim, their assumption that the *total* field external to the source is proportional to the radiation reaction is not warranted.

Hoyle and Narlikar's appeal to this argument is somewhat surprising because they point out that Wheeler and Feynman's derivation of the radiation reaction relies crucially on the condition of complete absorption: "The crucial issue is that of *complete absorption*.... The self-consistency argument [establishing that the absorber contributes a field $F_{rad} = 1/2(F^i_{ret} - F^i_{adv})$ acting on the source] will not work if the universe is an imperfect absorber" (Hoyle and Narlikar 1995, 118–119; italics in original).

Now, in fact there is a rather simple argument that can show there can be no asymmetry between retarded and advanced field representations in a theory with symmetric particle interactions. For a necessary condition for the adequacy of a fully retarded representation is that Wheeler and Feynman's absorber condition (6.6) holds. And from this it follows that a fully advanced representation is possible as well. Recall that the fully retarded representation of the total field acting on a source i is

$$\sum_{k \neq i} F^k_{ret} + \frac{1}{2}(F^i_{ret} - F^i_{adv}), \tag{6.8}$$

such that the field acting on some small test charge close to i is given by (6.9): $\sum_k F^k_{ret}$. That is, (6.8) guarantees that the radiation reaction has the correct sign

132 *Inconsistency, Asymmetry, and Non-Locality*

and that the field apparently associated with the source i is fully retarded. Since we assume that the total field is due to the symmetric half-retarded, half-advanced fields associated with all charges, we get

$$\frac{1}{2}\sum_{all\ k}(F_{ret}^k + F_{adv}^k) = \sum_{all\ k} F_{ret}^k, \quad (6.23)$$

which implies Wheeler and Feynman's condition

$$\sum_k (F_{ret}^k - F_{adv}^k) = 0. \quad (6.6)$$

But from (6.6), together with the fact that the total field can be written as $\frac{1}{2}\sum_{all\ k}(F_{ret}^k + F_{adv}^k)$, it follows that a fully advanced field representation of the field is possible as well.

Thus, there is a completely general argument for why any attempt to show, within the context of symmetric particle interactions, that a fully retarded field representation is uniquely picked out must fail. If a fully retarded representation is adequate, then necessarily so is a fully advanced representation. The only chance for breaking the symmetry between the retarded and advanced representations (6.8) and (6.10) is to argue that the retarded absorber field in (6.8) is zero, while the advanced absorber field in (6.10) is not. We have already seen that Wheeler and Feynman's statistical argument for this claim fails. But so far we have characterized absorbers only in a general formal way, embodied in equations (6.3) and (6.6). Can a more concrete physical model of the absorber perhaps provide the missing argument for the asymmetry? This is the question to which I want to turn next.

4. Absorbing Media with Complex Refractive Index

Wheeler and Feynman's detailed derivation of the radiation reaction due to an absorber assumes that the source is surrounded by a spherically symmetric absorber which is modeled as a dilute plasma of free charged particles. All the absorber particles are assumed to be sufficiently far from the source that the Coulomb interactions are negligible and only the radiation fields of source and absorber have an effect on each other. The derivation begins by asking what the response field of an individual absorber particle is at the location of the source due to the retarded component of the field of the source. This is calculated in two steps. First, we determine with the help of the Lorentz force law the acceleration of a typical absorber particle due to the field of the source, which is a retarded field and, since it is a radiation field, varies with R^{-1}. Then we calculate the advanced field of the absorber particle at the source, where we assume that the response of the absorber particle is half its advanced field. In chapter 2, we discussed this stepwise approach to modeling electromagnetic phenomena, which involves combining the effects of fields on the motion of a charge and the contribution of a charge to the field in, strictly speaking, inconsistent ways. Notice that Wheeler and Feynman assume that the field of the source affects the absorber particles according to the Lorentz force law, even though their aim is to derive a different equation of motion, the Lorentz–Dirac equation, for the source. This is an example of the kind of inconsistent modeling that is characteristic of classical electrodynamics and that we discussed in detail in chapter 2.

The resulting field depends on R^{-2}, where one factor of R^{-1} is contributed by the retarded field of the source and one by the advanced field of the absorber particle, and the two time lags associated with retarded and advanced fields cancel. That is, the advanced field associated with the absorber particle acts simultaneously with the source's acceleration. In order to determine the advanced response field of the entire absorber, one then introduces an expression for the particle density and integrates over the absorber region, which is taken to extend to spatial infinity. Two worries arise at this point. First, the response field of each absorber ion is proportional to the acceleration of the source and not to the derivative of the acceleration, as would be required for the theory to generate the radiation reaction term of the Lorentz–Dirac equation. Second, the resulting integral does not converge (because the R^{-2} dependence of the response field associated with individual absorber particles is canceled by the radial integration factor $R^2 dR$).

Wheeler and Feynman argue, however, that the derivation so far has ignored the effect of the absorber on the retarded field of the source. The field of the source will set the plasma ions into motion, and they in turn will radiate. But since the medium is acting as an absorber, not all the energy imparted on the plasma ions will be reradiated, and some of the energy will be converted into thermal energy through collisions between plasma ions (with the effect that the entropy of the plasma will increase). Phenomenologically, the effect of the absorber on the field can be characterized in terms of a complex refractive index associated with the medium. Taking the refractive index into account alters the derivation of the response wave in just the right ways to address the two worries. First, the real part of the refractive index will introduce a phase lag for the retarded field compared with propagation in vacuum, and this helps to ensure that the radiation reaction has the right dependence (on the derivative of the acceleration). And, second, the imaginary part, which is negative, results in a damping factor.

Putting it all together, Wheeler and Feynman show that if we assume that the field acting on the absorber is proportional to the fully retarded field of the source—that is, if, as they put it, the absorber experiences a field "?" times the fully retarded field—then the response of the absorber will be proportional to half the difference between the sources of retarded and advanced fields, with the same factor of proportionality "?". What remains is to determine that factor. Wheeler and Feynman argue that the total field diverging from the source and acting on the absorber has two components, the first given by the original retarded field associated with the source (which, according to the assumption of a symmetric interaction, is 1/2 the retarded field) and the second calculated to be ?/2 times the retarded field. Thus, the proportionality factor has to satisfy the following equation: $? = 1/2 + ?/2$, from which it follows that $? = 1$.[5]

One might object to this derivation by arguing that Wheeler and Feynman's argument only shows that the field acting on a test particle close to the source will be fully retarded, since, as they point out, a test particle will not be able to tell whether the field is due to a source or to the absorber. But, one might say, this argument does not show that the field acting on the *absorber* is equivalent to the fully retarded field of the source. For there is an important difference between the absorber and a test charge: Unlike a test charge, absorber particles *can* 'tell' what the source of the field components is, since in the action-at-a-distance theory no particle interacts with its own field. Thus, should one assume that the field acting on the absorber is only 1/2 the retarded field associated with the source?[6] Yet each absorber particle j will interact with the fields of all other absorber particles $k \neq j$, and the mistake we make in including the field of each individual absorber particle in the total field acting on that particle is negligible.

The complex refractive index introduces a temporal asymmetry into the theory. Can we account for the asymmetry of radiation with the help of this asymmetry? According to Davies, the answer is 'yes.'

> The advanced self-consistent solution, which is allowed on purely electrodynamic grounds, is ... ruled out as being overwhelmingly improbable, because it would require the cooperative "anti-damping" of all the particles in the cavity wall, corresponding to a positive imaginary part [of the refractive index], i.e. exponentially growing disturbance. Ions would become collisionally excited, and radiate at the precise moment throughout the wall to produce a coherent converging wave to collapse onto the cavity centre at just the moment that the charged particle there was accelerated. (Davies 1974, 144)

Davies's appeal to the improbability of advanced fields here might be taken simply to be a repetition of Wheeler and Feynman's statistical argument. Yet his direct appeal to the complex refractive index also suggests a somewhat different explanation of the asymmetry. In an absorber characterized by a refractive index with a negative imaginary part, fields are (invariably or at least almost always) damped in the normal, thermodynamic time direction. This, Davies suggests, means that no nonzero field can exit the absorber at its future boundary. For a nonzero field at the boundary would require exponentially growing disturbances within the absorber, which are incompatible with a negative imaginary refractive index. By contrast, the refractive index is clearly not incompatible with nonzero fields entering the future absorber from the past. Thus, there is an asymmetry between the retarded and advanced field representations (as given, for example, in (6.16)): Since there is no nonzero field at the future boundary of the past absorber, the retarded absorber field is zero, while the advanced absorber field is not zero. The field in the absorber cavity is equal to the fully retarded field of the source at its center, while the advanced representation includes a nonzero advanced field from the future absorber.

Davies does not make this explicit, but his argument depends on the fact that the absorber is ideal. For clearly, if the past absorber were nonideal, a nonzero field in the future of the absorber could be associated with a very strong free field in the past of the absorber that is damped as it travels through the absorber. There would have to be no anti-thermodynamic growing disturbances within the absorber. The problem with the case of an ideal absorber is that any finite field exiting the absorber would have to be accompanied by a field that is infinite in the absorber's past. Thus, nonzero fields at the future boundary of the past absorber are excluded

Absorber and Entropy Theories of Radiation 135

because they would have to be associated *either* with growing disturbances within the absorber, which are excluded on thermodynamic grounds, *or* with infinite fields in the past of the absorber, which are unphysical.

As I have presented it, Davies's explanation of the asymmetry is distinct from Wheeler and Feynman's statistical argument in that Davies invokes a phenomenological thermodynamic asymmetry as premise. The account is superior to the statistical argument, since it does not seem to be guilty of a temporal double standard. Rather, Davies explains the asymmetry by appealing to an explicitly time-asymmetric premise. Nevertheless, I believe that the argument ultimately fails to provide a satisfactory solution to the puzzle of the arrow of radiation, since a symmetric interaction between charged particles is incompatible with Wheeler and Feynman's physical model of an absorber. Since this criticism affects not only Davies's explanation but also Wheeler and Feynman's theory more generally, I want to develop it with some care. In the end I want to argue that there is a fundamental flaw in the detailed derivation of the response wave given by Wheeler and Feynman. But in order to motivate my argument, I want to suggest a number of smaller, related worries first.

Wheeler and Feynman's detailed derivation of the response field of the absorber focuses only on the future absorber. As we have seen, they argue that the retarded component of the field of the source results in a response of the absorber that is equal to $1/2(F^i_{ret} - F^i_{adv})$, thus canceling the advanced field of the source and doubling its retarded field to full strength. But what if we had begun the derivation with the past absorber instead? The worry is that then the refractive index of the absorber has the opposite effect from the one that is desired: Instead of ensuring that the integral over response fields of all the absorber particles converges, the advanced field of the source would be anti-damped as it traveled backward in time through the absorber and the integral would diverge. (This can easily be seen if we look at the time-reversed situation: a retarded source field entering an anti-absorber.) Now, of course it is a consequence of the derivation involving the future absorber that the advanced field of the source is canceled by the absorber response field, but how do we know that the future absorber is the correct place to start?

The answer we can extract from Davies's explanation of the asymmetry seems to be that the very fact that the fields diverge in the derivation starting with the past absorber tells us that this derivation cannot lead to a physically acceptable solution. A different reply is given by Hogarth (who, as we have seen, also wants to allow for nonideal absorbers). Hogarth simply reverses the sign of the imaginary part of the refractive index for past absorbers to ensure that the past integrals converge as well! This means, of course, that Hogarth's past 'absorber' is what we would normally call an *anti-absorber*, since fields propagating in it in the forward-time sense are anti-damped due to the positive imaginary refractive index. In particular, on Hogarth's account the advanced wave associated with the source is zero in the remote past, then gathers strength as it passes through the past absorber to exit the absorber at full strength and converge into the source. Note that this 'trick' is incompatible with Davies's solution, since if we reverse the sign of the imaginary refractive index for the past absorber, there is once again a complete symmetry between past and future absorbers: In both cases fields are damped in the temporal direction away from the source.

On first sight Hogarth's proposal ought to strike one as rather surprising. How can it be that the absorber medium (which, after all, is *one and the same* medium that absorbs radiation in the past and in the future) is absorbing for retarded radiation and anti-absorbing for advanced radiation? To be sure, Hogarth argues that due to the expansion of the universe, there are differences in the wavelengths of the radiation and in the particle densities between the two cases. Still, one might doubt that there is any realistic physical process which could account for the reversal of the direction of absorption and also whether the proposal would not require making what appears to be an incoherent assumption—that entropy has to increase in the medium in both time directions. Moreover, one might worry that, as Davies (1975) and the mysterious Mr. X (Gold and Bondi 1967, 13) have argued, thermodynamic considerations are smuggled into Hogarth's cosmological reduction through the different signs in the imaginary refractive index. Hogarth himself sums up this criticism up as follows (and seems to willing to concede the point, absent some other argument for the convergence of the integrals): "I think that X's criticism is that I need to make certain integrals convergent and that I picked up a little piece of thermodynamics and included it for convenience" (Gold and Bondi 1967, 24).

In fact, however, if we assume that each charge physically contributes a symmetric field—as the Wheeler–Feynman theory does—then Hogarth's assumption that the two absorbing regions are characterized by two different phenomenological thermodynamic arrows is far more compelling than Davies's hypothesis that the absorber is uniformly absorbing in the future-time sense. Indeed, far from being nothing but a desperate mathematical trick, Hogarth's conception of an 'absorber' characterized by an imaginary refractive index whose sign changes depending on whether retarded or advanced radiation enters the absorber appears to be correct from the standpoint of statistical physics.

If we model the absorber as a large spherical shell with a briefly accelerating source somewhere near its center and, once again, assume that the source is physically associated with a symmetric half-retarded, half advanced field, then we will find that there are coherent oscillations among absorber particles on the inner surface of the shell both before and after the source accelerates, and that coordinated motions and the associated fields are damped toward the outside of the shell. Microscopically, the absorber is governed by a time-symmetric dynamics, and any asymmetry of the absorber is entirely due to an asymmetry of initial or final conditions: If the retarded or advanced fields set up correlations among the absorber particles, then these correlations will be damped in the direction away from the source and the medium will act as absorber in the presence of retarded radiation, and as anti-absorber in the presence of advanced radiation. Coherent oscillations will partly be associated with waves diverging from the source in the center and will partly appear as 'spontaneously' arising coordinated motions among absorber particles which give rise to waves collapsing inward toward the source. The former are due to retarded fields of the source and the latter are due to its advanced fields.

In either case the fields of the source lead to low-entropy states of the absorber, which evolve into higher states. In the former case the field results in a low-entropy *initial* state, which evolves into a higher state in the future. In the latter case the field results in a low-entropy *final* state, which evolved from a higher state with higher entropy in the past. The local entropy minimum as the coherent oscillations arise, leading to a converging wave, is explained by that converging wave — the advanced field of the source. The absorber is not an isolated system but is coupled to the advanced field of the source, which is responsible for the local entropy minimum of the absorber. Of course this is simply the time reverse of the decrease in entropy due to the retarded field. In the retarded case the *decrease* in entropy is explained by the coupling of the absorber to the retarded field of the source, and the subsequent entropy *increase* is explained statistically. In the advanced case the *decrease* of entropy is explained statistically, while the subsequent *increase* of entropy is explained by the coupling of the absorber to the advanced field of the source. This means that, as Hogarth assumes, the direction of entropy increase governed by statistical reasons depends on whether the interior wall of the absorber is excited by retarded or by advanced radiation. But this is what we would expect, given that the underlying dynamics is time-symmetric: Entropy increases in both time directions away from local entropy minima.

I now want to return to my main criticism of Wheeler and Feynman's absorber model. So far I have not questioned whether Wheeler and Feynman (as well as Davies, Hogarth, and Hoyle and Narlikar) are in fact correct in claiming that the future absorber with negative imaginary refractive index leads to a finite absorber response wave in the manner sketched above. But this claim is false. Let us return to the very beginning of these derivations. Wheeler and Feynman begin by asking, first, what the acceleration of a typical absorber particle is, due to the field of the source, and, second, what the field force associated with an absorber particle is, at or near the source. After writing down what the strength of these interactions would be in the absence of any other charges, they then correct the field of the source by taking into account the refractive index of the absorber medium. Yet, curiously, the advanced field associated with the absorber particle is not similarly adjusted. As Wheeler and Feynman explicitly say:

> The advanced force acting on the source due to the motion of a typical particle of the absorber is an elementary interaction between two charges, propagated with the speed of light in vacuum. On the other hand, the disturbance which travels outward from the source and determines the motion of the particle in question is made up not only of the charge, but also of the secondary fields generated in the material of the absorber. (Wheeler and Feynman 1945, 161)

But there is no justification for this difference in treatment. Why should the advanced field of absorber particles be an interaction in a vacuum and be unaffected by the presence of other absorber particles, when the retarded field of the source interacts with other absorber particles on the way to an absorber particle in the interior of the absorber? If the retarded field of the source interacts with other absorber particles on its way to the particle generating a response wave, then the advanced field of an absorber particle should likewise interact with other absorber particles on its way back to the source. But if the advanced field of an absorber particle is adjusted similarly to the retarded field of the source, then there should be not only one phase factor in the absorber response wave but two: a first factor accounting for the damping of the retarded field of the source, and a second factor accounting for the effect of the medium on the advanced field associated with an absorber particle. The trouble is, however, that this second phase factor should enter with the opposite sign. Since the retarded field of the source is damped in the absorber, the advanced field of a typical particle will be anti-damped in the backward time direction.

We can think of it this way. All fields are damped in the absorber in the forward-time sense, including the advanced field associated with a typical absorber particle. That is, the field converging on an absorber particle has to be larger far away from the source than it would be in a vacuum, if it is to have the right strength as it collapses into the particle after having traveled through the absorber. Compare the behavior of advanced fields in the absorber with that of advanced fields in Hogarth's anti-absorber: In the anti-absorber the advanced field of the source is damped in the direction away from the source with which it is associated—that is, toward the past—which appears to ensure that the integral over the response fields of all the absorber particles converges. By the same token, advanced fields are also anti-damped in an absorber in the past direction.

The result is that the two phase factors, as well as the damping and anti-damping factors, cancel each other and the integral over all absorber particles no longer converges. For the integral to converge, the retarded field of the source has to fall off faster than R^{-1}. That is, the acceleration of each absorber particle has to be less than it would be in a vacuum. This is achieved by introducing an exponential damping factor $e^{-\alpha R}$ due to the complex refractive index of the absorber. But while this alone would ensure that the acceleration falls off fast enough with R, the advanced field of each absorber particle ought to be anti-damped, contributing a factor of $e^{\alpha R}$ that is just big enough to cancel the factor necessary for convergence.

Thus, Wheeler and Feynman's detailed derivation of the absorber response wave contains a serious and, to my mind, fatal flaw in that it only selectively takes account of the effect of the absorber on the fields propagating in it. If, on the one hand, we ignore the damping factor associated with the absorber, the absorber response field diverges. Yet if, on the other hand, we take the presence of the absorber *fully* into account, the response field diverges as well. What this suggests is that a temporally asymmetric conception of an absorber with complex refractive index is incompatible with Wheeler and Feynman's proposal of a symmetric interaction between charged particles.[7]

This ends my survey of attempts to account for the asymmetry of radiation strictly within the framework of Wheeler and Feynman's absorber theory with time-symmetric particle interactions. I have argued that neither Wheeler and Feynman's formal absorber model nor their physical model of absorbing media can provide an adequate explanation of the asymmetry of radiation, if we assume that charged particles are physically associated with a time-symmetric field. I now want to examine a possible alternative way in which certain aspects of Wheeler and Feynman's answer might be developed into a solution to the puzzle—Price's proposed "reinterpretation" of the mathematical core of the Wheeler–Feynman theory.

5. Price's Argument for the Microsymmetry of Radiation

In the previous chapter I distinguished two putative asymmetries associated with radiative phenomena:[8]

(*i*) (RADASYM) There are many situations in which the total field can be represented as being approximately equal to the sum of the retarded fields (and not of the advanced fields) associated with a small number of charges.

and

(*ii*) All accelerated charges (or sources) are physically associated with fully retarded (but not with fully advanced) radiation fields.

Both these asymmetries, on the standard view, can be exhibited by microscopic fields associated with individual charged particles. By contrast, Huw Price argues that the apparent asymmetry of radiation arises only for fields of macroscopic collections of charges and maintains that the asymmetry should properly be characterized by the claim that

(*iii*) Organized waves are emitted, but only disorganized waves are absorbed.[9]

Price emphasizes the distinction between emitters and absorbers in his account. An emitter is a charge or collection of charges that emits electromagnetic energy, while an absorber is a charge that absorbs energy. Since Price holds that only emitters of radiation are associated with retarded waves, he contrasts (*iii*) not with (*ii*) but rather with

(*iv*) All emitters produce *retarded* rather than *advanced* wave fronts.

The difference between (*ii*) and (*iv*) is that according to (*ii*), *all* electric charges, independently of whether they act as emitters or absorbers of energy, are associated with retarded fields, while (*iv*) makes a claim only about charges that act as net emitters. Price does not discuss the asymmetry (*i*).

Price argues, first, that the truth of (*iv*) does not imply that radiation is asymmetric and, hence, that the putative asymmetry of radiation is best captured by (*iii*), and not by (*iv*). He argues, second, that (*iii*) is false on the micro level, where radiation is fully symmetric. In his argument for the microsymmetry of radiation, Price appeals to a reinterpretation of Wheeler and Feynman's infinite absorber theory, which he takes to show that, for any given configuration of emitters and absorbers, the retarded fields associated with the emitters can equivalently be represented as a superposition of coherent converging (advanced) waves centered on the absorber particles. From this Price concludes that radiative processes are symmetric on the micro level in the sense that both emissions and absorptions can be associated with organized waves. That is, on the micro level (*iii*) is false and the following holds:

(v) Both emitters and absorbers are centered on coherent wave fronts (these being outgoing in the first case and incoming in the second).

Price believes that the puzzle of the arrow of radiation presents a genuine puzzle; he believes that *if* electromagnetic radiation is asymmetric, then this calls out for an explanation. But he argues that the antecedent of this conditional is false on the micro level—radiation is symmetric on that level—and that therefore only the apparent macro asymmetry of radiation is in need of an explanation. The apparent asymmetry of radiation on the macro level, according to Price, is due to the fact that, because of cosmological initial conditions, there are large, macroscopic coherent emitters but no macroscopic *coherent* absorbers. Thus, Price believes that the asymmetry of radiation has the same status and the same source as the asymmetry of thermodynamics. Both are more de facto asymmetries due to asymmetric cosmological initial conditions.

I believe that Price's account faces a number of problems. First, his reinterpretation of the Wheeler–Feynman theory is inconsistent both with the central assumption of that theory and with classical electrodynamics in general. All representations of a field—retarded, advanced, or some combination—have to include wavelets centered on all charges present. This consequence of the Maxwell equations is violated by Price's proposal. (For detailed arguments for these claims, see Frisch 2000). Second, (v), properly understood, follows straightforwardly from the Maxwell equations and, contrary to what Price seems to think, is not in need of support from a reinterpretation of Wheeler and Feynman's theory. Yet, third, the truth of (v) does not establish Price's conclusion that radiation is symmetric on the micro level, since the asymmetry of radiation is correctly captured by (i) or (ii), but not by (iii); and (v) does not imply that (ii) is false. Thus, Price has not shown that electromagnetic radiation is symmetric on the micro level and therefore has not solved the puzzle of the arrow of radiation. Finally, the fact that Price's account does not engage with (i) is problematic. As phenomena such as synchrotron radiation seem to suggest, the asymmetry between initial and final conditions governs microscopic as well as macroscopic phenomena. Price's symmetric condition (v) cannot account for that fact.

Price proposes his reinterpretation of Wheeler and Feynman's infinite absorber theory as an argument for (v), the claim that "[b]oth emitters and absorbers are centered on coherent wave fronts (these being outgoing in the first case and incoming in the second)." Even though this reinterpretation is deeply problematic, (v)—or, more precisely, the part of the claim not in parentheses—is true in classical electrodynamics. The claim follows directly from the Maxwell equations and is in no need of support from a controversial theory such as that of Wheeler and Feynman.

There are two claims implicit in (v). The first is that all sources, whether they act as emitters or absorbers of radiative energy, are centered on coherent wave fronts. The second claim is that emitters are associated with outgoing waves, while absorbers are associated with incoming waves. The second claim can be read as a conditional: If all sources are centered on coherent wave fronts, then emitters are associated with outgoing wave fronts and absorbers are associated with incoming wave fronts. Now, the first claim is an immediate consequence of the Maxwell equations, according to which *every* charge, be it a net absorber or a net emitter of energy, contributes a component to the total field which is centered at the source (where that field component can be either retarded or advanced, depending on the particular representation chosen). What about the second part?

In light of our discussion in the previous chapter, we can see that the notion of a field being *associated* with a charge is ambiguous. Recall our distinction between the total fields and the field physically associated with a charge. If we are interested in the total field and represent that field in terms of an initial- or final-value problem, then it is irrelevant whether a charge acts as net emitter or net absorber of energy. In an initial-value problem all charges are mathematically associated with retarded fields, independently of whether they act as sources or as sinks. Similarly, in a final-value problem all charges are mathematically associated with advanced fields. Thus, if Price's claim is that emitters *can* be mathematically associated with retarded waves, and absorbers with advanced waves, then this claim is true, and follows simply from the facts about initial- and final-value representations that we rehearsed in chapter 5.

What if, alternatively, we read Price as claiming that absorbers are *physically* associated with advanced fields? The strict identification of absorption processes with advanced waves and of emissions with retarded waves has some intuitive appeal. Given a specific temporal orientation, the retarded solution to the wave equation describes a disturbance that originates at the source at a time t_0 and travels outward for times $t > t_0$, while the advanced solution describes a disturbance that converges into the source at times $t < t_0$ and 'disappears' at the source at time t_0. Intuitively, the former solution seems to characterize an emission process and the latter solution an absorption process. However, as Zeh (2001, 16) argues, whether charges are physically associated with retarded or advanced fields is independent of whether the charge acts as an absorber or an emitter. For a charge can act as an absorber even if by hypothesis it is taken to be physically associated with a retarded field. Zeh considers the case of an incoming field that interacts with a source which in turn emits a purely retarded field, where the retarded field interferes destructively with the incoming field. Energy then flows from the field into the source, which, therefore, acts as an absorber, even though its physical contribution to the total field is a purely retarded field. Similarly, the emission of energy can be associated with a purely advanced field. Thus, I propose that we read Price's claim as being concerned with possible *mathematical* representations of the field. Then it is true that emitters and absorbers can be associated with diverging and coherent waves, respectively, but this fact has nothing to do with whether the charges in question act as net emitters or absorbers of energy. That is, Price is right in saying that on the micro level (v) is true (if correctly understood). (v) is true, but does that mean electromagnetic radiation is symmetric on the micro level?

There are mathematical representations of the total field in which net emitters of radiation are associated with purely retarded fields; and there are mathematical representations of the total field in which net absorbers of radiation are associated with purely advanced fields. But what are the relations between this claim and the two putative asymmetries associated with radiation that I distinguished in the previous chapter, claims (i) and (ii) above? Price argues that, given that radiation is symmetric in the sense captured by (v), the fact that emissions can be represented in terms of purely retarded fields does not imply that radiation is asymmetric in any interesting sense. I take it that his banking analogy is meant to provide an argument to that effect (Price 1996, 58–61). According to Price, since deposits into a bank account are temporal inverses of withdrawals, there is nothing temporally asymmetric about banking as a whole. Even if *deposits* were in some sense temporally asymmetric, *banking as a whole* is temporally symmetric: Transactions that look like deposits in one temporal direction turn into withdrawals if the direction of time is reversed, and withdrawals turn into deposits. Similarly, Price holds, electromagnetic absorptions can be construed as temporal inverses of emissions and, thus, radiation processes as a whole are not temporally asymmetric.

I will argue in the next chapter that despite the symmetry in *mathematical* representations, the theory is asymmetric in that accelerated charges are *physically* associated with fully retarded (but not with fully advanced) radiation fields. That is, the truth of (ii) is compatible with (v). For now, however, I want to focus on the asymmetry expressed by (i), according to which incoming fields, but not outgoing fields, are approximately equal to zero in many situations. Applied to Price's banking analogy, the latter asymmetry would consist in the fact that while for most initial times of interest to us the balance in our bank account was zero, fortunately at final times the balance was usually positive. As it stands, Price's theory does not appear to have the resources to account for this asymmetry. Yet, whatever else we might take the asymmetry of radiation to consist in, the most obvious and least controversial asymmetry is that between initial and final condition embodied in (RADASYM). Any satisfactory account of radiation phenomena ought to provide an account of that asymmetry.

One might think that Price's appeal to an asymmetry in the macroscopic initial conditions of the universe can help explain this asymmetry. Price claims that there is an asymmetry at the macroscopic level between emitters and absorbers: "Large-scale sources of coherent radiation are common but large receivers, or 'sinks,' of coherent radiation are unknown" (Price 1996, 71). And one might hope that this de facto asymmetry can play a role similar to the one absorbers are intended to play in the more traditional absorber theories we discussed above. However, Price's notion of a large-scale absorber or receiver of radiation is distinct from the notion of an absorber in the tradition of Wheeler and Feynman. For the argument Price is offering is not meant to be "dependent on the thermodynamic properties of the absorber" (Price 1996, 72). And Price explicitly denies that Wheeler and Feynman's absorber plays any role in his argument (ibid.). More important, there simply are no large-scale absorbers or sinks of radiation, according to Price, which might somehow account for the fact that incoming fields are zero. Thus, as far as I can tell, Price's account cannot provide an explanation of the fact that incoming fields, but not outgoing fields, can generally be set equal to zero.

6. Conclusion

In this chapter I criticized absorber accounts of the radiation asymmetry in the tradition of Wheeler and Feynman's time-symmetric action-at-a-distance theory and what one might call 'pure entropy accounts' of the asymmetry. Wheeler and Feynman appeal to two distinct conceptions of an absorber: a general formal model and a specific physical model of an absorber as dilute plasma. I argued that all attempts to derive the asymmetry within the formal model by invoking statistical or cosmological arguments fail. Indeed, I argued that they must fail, for the formal absorber model implies that radiation must be completely symmetric: If there is a fully retarded representation of the field, there is a fully advanced representation as well. But attempts to derive an asymmetry within the specific absorber model fail as well, since, as I argued, the model is in fact incompatible with the assumption of a symmetric half-advanced, half-retarded field. Wheeler and Feynman (and others) can arrive at well-defined results only by selectively ignoring the interaction between absorber and half of the field associated with a charge. Moreover, Price's reinterpretation of the Wheeler–Feynman theory is unsuccessful as well.

7

The Retardation Condition

1. Introduction

In the previous two chapters I examined several purported solutions to the puzzle of the arrow of radiation. Quite generally the electromagnetic field in a given space-time region can equivalently be represented as a sum of free incoming fields and retarded (or diverging) fields, or as a sum of free outgoing and advanced (or converging) fields. The puzzle associated with radiation fields is Why can radiation usually be represented as sum of purely retarded fields (with zero incoming fields) but not as sum of purely advanced fields? In other words, why can the total field generally be represented purely as superposition of divergent waves associated with the charged particles in question, but not purely as superposition of converging waves? All the answers we discussed were unsatisfactory.

In this chapter I want to defend what I take to be the most successful answer to this puzzle—an answer appealing to an explicitly time-asymmetric constraint that is best thought of as causal. In order to motivate the account, I want to begin with a discussion of H. Dieter Zeh's account of the asymmetry; Zeh also adopts Wheeler and Feynman's specific model for an absorber but, unlike the proposals in that tradition which we examined in the last chapter, makes no assumptions about the field physically associated with a charged particle. This account is, to my mind, the most promising proposal for reducing the radiation arrow to some other physical asymmetry, in that it is the only such account that actually allows us to derive the explanandum from the explanation proffered in a noncircular or question-begging way. Moreover, Zeh appears to be correct in arguing for the need to appeal to past absorbers in accounting for the asymmetry.

Nevertheless, I want to raise several worries about Zeh's account. First, it is unclear whether the account can explain an asymmetry more broadly than that captured in RADASYM. Second, Zeh's answer to the puzzle is in fact incompatible with the notion of a field physically associated with a charge. For if there is such a thing as a field physically contributed by a charge, then Zeh's answer is circular and needs to presuppose what it is intended to show: that radiation is fully retarded. And third, his explanation accounts for what appear to be thermodynamically probable initial conditions (i.e., weak, fully thermalized fields) by having to appeal to radically more improbable earlier initial conditions. As we will see, the account has to posit inexplicable correlations as part of the state of the early universe—correlations that are readily explained by the causal account I want to defend.

In section 3, I will present my solution to the problem in terms of the retardation condition. In section 4, I will try to dispel worries one might have about the legitimacy of the notion of a field associated with a charge and argue further that excluding this notion from the theory unduly restricts its content. I will also offer some reasons why we should think of the retardation condition as causal. In section 5, I will connect my account to recent interventionist and manipulability accounts of causation.

2. Zeh's Ideal Absorber

Zeh (2001)[1] argues that the asymmetry of radiative phenomena is due to ideal absorbers in the past of space-time regions in which we are interested and that these ensure that there are no source-free incoming fields. Zeh's definition of an absorber is equivalent to Wheeler and Feynman's specific model of an absorber as a medium characterized by a complex refractive index but, contrary to Wheeler and Feynman, Zeh makes no assumptions about the physical contribution of a charge to the field.

More specifically, Zeh defines an absorber as follows (where expressions in parentheses refer to the ideal case at a temperature of absolute zero): "A spacetime region is called '(ideally) absorbing' if any radiation propagating in it (immediately) reaches thermodynamical equilibrium at the absorber temperature $T (=0)$" (Zeh 2001, 22). This definition implies, according to Zeh, "that no radiation can propagate within ideal absorbers, and in particular that no radiation may *leave* the absorbing region (along forward light cones)" (ibid., p. 23; italics in original). Zeh says that laboratories prior to an experiment closely approximate an ideal absorber, and also offers some cosmological reasons why the past of our expanding universe constitutes a (nearly) ideal absorber.

Thus, Zeh's account is beautifully simple: The fact that incoming fields can be chosen to be approximately equal to zero in situations in which we are interested, simply follows from the fact that there are absorbers in the past of the relevant spatiotemporal regions. This account is in direct conflict with the cosmological explanation of the asymmetry proposed by Hogarth and by Hoyle and Narlikar, who claim, as we have seen in the last chapter, that the asymmetry can be accounted for by the fact that the *future* of the universe is ideally absorbing independently of its past, and argue that an ideal *past* absorber would not privilege a retarded field representation. By contrast, Zeh appeals to the purported fact that the *past* is ideally absorbing to account for the condition of zero incoming fields.

Zeh's explanation is similar in certain respects to Davies's argument for the asymmetry. Like Davies, Zeh needs to make the additional assumption that fields have to be finite everywhere and can exclude nonzero advanced fields in the immediate future of the absorber only if the absorber is ideal. (For in the case of a nonideal absorber there could be nonzero fields on its future boundary due to very strong fields in the absorber's past.) If the absorber is ideal in Zeh's sense, then incoming fields are damped out instantaneously. This means that the imaginary part of the refractive index characterizing the absorber is infinite, and that any advanced field in the future of the absorber would have to be singular at the absorber boundary. The crucial difference between Zeh's and Davies's accounts is that Davies relies on a symmetric particle interaction, which generates the correct result only under the assumption that the future absorber contributes an advanced response wave that destructively interferes with the advanced field of the source in question. And as we have seen in the last chapter, this assumption relies on the unwarranted further assumption—needed to ensure that the future absorber makes only a finite contribution to the field—that only the retarded field, and not individual advanced response waves, interacts with the absorber. Thus I think Zeh's solution to the puzzle of the arrow of radiation is successful in ways in which Davies's is not. More precisely, Zeh succeeds in showing the following: If the past acts as an ideal absorber and the electromagnetic field is constrained to being finite everywhere, then, if we choose as initial-value surface a hypersurface at (or in the immediate future of) the absorber boundary, the field in the future of that surface is fully retarded.

Nevertheless, I do not believe that Zeh provides a completely satisfactory solution to the puzzle of the arrow of radiation. The first worry I have concerning Zeh's explanation concerns its scope. In chapter 5, I argued that (RADSYM) delineates a reasonably precisely characterizable class of phenomena which exhibit a temporal asymmetry, but that these phenomena do not exhaust the phenomena exhibiting the radiation asymmetry. Any successful explanation of the asymmetry would have to be broad enough to extend to other, perhaps less precisely characterizable, phenomena as well. An example of a phenomenon not covered by RADASYM is this. Imagine a radiating source in the presence of an approximately constant incoming field of half the amplitude of the radiation. In this case there would be a clearly identifiable diverging field, and we would not expect to observe the temporal inverse of a converging wave superposed over a large constant field. It is not clear how Zeh's account might handle such a case. The asymmetry here is that there are situations (or that it is easy to set up situations) where the incoming field is approximately constant and the outgoing field is a superposition of a constant field and a diverging wave, but that the time reverse of such situations does not occur or is extremely difficult to set up. Since the incoming field is not zero, we know that there cannot be an ideal absorber in the immediate past of any initial-value surface we might choose to represent this case.

My second worry concerns the status of the notion of a field associated with a charge in Zeh's account. In chapter 5, I distinguished two putative temporal asymmetries of electromagnetic fields: an asymmetry of the *total* fields, which generally satisfies the Sommerfeld radiation condition of $F_{in} = 0$, and an asymmetry of the fields physically associated with each charged particle. The view presented in many textbooks on electrodynamics is that fields associated with charged particles are fully retarded—that is, they are asymmetric in the latter sense—and that this asymmetry helps to explain the former asymmetry: It is because each charge contributes a fully retarded field to the total field that incoming fields, but not outgoing fields, can usually be set to zero in situations in which we are interested.

Advocates of ideal absorber theories in the tradition of Wheeler and Feynman also assume that it makes sense to speak of the field physically associated with a charge—with the difference, of course, that they suppose that the field force associated with each charge is time-symmetric. This view is explained by Hogarth, who—after pointing out that the most general solution to the field equations for a single source can be written as a linear combination of retarded and advanced fields, $F = bF_{ret} + (1-b)F_{adv}$—maintains that the only alternative to "arbitrarily" introducing a time asymmetry is to choose the symmetric solution with $b = 1/2$. Hogarth argues that a "classical theory of electrodynamics...cannot be considered complete without somehow specifying the value of b," and that "the assumption that the physically significant solutions of the field equations should not in themselves introduce an arrow of time leads directly to the unique value of 1/2 for b" (Hogarth 1962, 366).

Zeh disagrees with this line of reasoning. That the theory does not by itself introduce the asymmetry, on his account, does not force us to postulate that the field associated with a charge is symmetric. In fact, he appears to hold that there is no need at all for the theory to specify any field associated with a charge *independently of particular initial or final conditions*. On the one hand, once initial or final conditions are given, the value of b is automatically fixed. The field equations are differential equations that define a Cauchy problem, which can be solved given specific initial or final conditions. In the case of an initial-value problem, the field mathematically associated with each charge is fully retarded, which means that $b = 1$. In the case of a final-value problem, the field mathematically associated with each charge is fully advanced and $b = 0$. On the other hand, without specific initial or final values we would not expect the theory to provide a determinate value for b. In fact, on Zeh's conception it does not even make sense to ask what the value of b is independently of particular initial or final conditions, since a differential equation like the field equation has a determinate solution only for a set of specific initial (or final) values. There is no such thing as the field associated with a charge independently of a specific initial- or final-value problem.

The notion of a field physically associated with a charge is in fact incompatible with Zeh's explanation of the asymmetry of radiation. For if each charge is physically responsible for a certain component of the total field, Zeh's account cannot explain why this component does not include an advanced part. If Hogarth's question as to what the field of a single source is, has an answer, then Zeh's explanation of why fields are fully retarded is question-begging.

The temporal asymmetry of Zeh's absorber is a thermodynamic asymmetry. As in the case of Wheeler and Feynman's absorber, the absorbing medium can be modeled as a dilute plasma of free charged particles. Particles in the plasma constituting the absorber are accelerated by the incoming radiation. But the particles do not reradiate the incoming radiation coherently; rather, the incoming energy is spread among many degrees of freedom through random collisions among the plasma particles. Thus, as the plasma absorbs radiation, its entropy increases. The question is how this time asymmetry in the entropy can be accounted for by microphysical statistical considerations. The thermodynamic arrow is commonly explained by an appeal to an asymmetry between initial and final conditions. That entropy increases (for the most part) is explained, roughly, by arguing that transitions from highly improbable low-entropy states to much more probable high-entropy states are overwhelmingly more likely than transitions from low-entropy states to other low-entropy states. A well-known objection to explaining the thermodynamic arrow in this manner is Loschmidt's reversibility objection which argues that if the underlying dynamics is time-reversal-invariant, then for each trajectory in phase space that carries an improbable state into a probable state, there is a trajectory that carries a probable state into an improbable state. Thus, entropy increases should be as likely as entropy decreases. A standard reply to this objection is to argue that the initial state of the universe was one of extremely low entropy and that, hence, initial states of quasi-isolated subsystems have very low entropy as well. And it is this de facto asymmetry between initial and final states of the universe or of its subsystems that can explain why entropy increases (in the normal time direction).

Applied to the absorber, the most probable macro state of the absorber is one in which the plasma ions are in random motion. A much less probable, low-entropy state is one in which plasma ions oscillate coherently (and, hence, radiate coherently). Now an isolated plasma would be overwhelmingly likely to remain in a high-entropy state. But if ions at the plasma boundary are set into coherent motion by incoming radiation fields, then it is most likely that the entropy of the plasma will increase due to collisions between plasma ions, which have the effect of spreading the energy of the incoming radiation among many different degrees of freedom. Due to random collisions among the plasma ions, not all the energy imparted to the ions by the incoming radiation will be reradiated. Thus the medium acts as an absorber, which phenomenologically can be described in terms of a complex refractive index with a negative imaginary part.

The underlying dynamics governing the plasma particles is time-symmetric, however. The plasma will act as an absorber (instead of as an anti-absorber) only if there is no process that can lead to a low-entropy final state, which (running the dynamics backward) would have to have evolved from a high-entropy initial state. But now it should be obvious that in trying to provide a microphysical basis for the complex refractive index characterizing the absorber medium, we cannot allow for a notion of the field physically contributed by a charge. If there were such a field, Zeh would have to assume that it is fully retarded. Fully retarded fields associated with sources in the absorber's past would be responsible for coherent fields at the absorber's past boundary that are damped in the forward time direction. Similarly, if the fields associated with charges contained an advanced component, then these fields would provide a process that could result in coherent motions among absorber particles on its future boundary and, hence, in a low-entropy final state! The absorber would then act as an 'anti-absorber' damping the advanced radiation toward the past time direction. Thus, if there are fields physically contributed by each charged particle, Zeh has to assume that these fields are fully retarded in order to ensure that the plasma acts as an absorber rather than an anti-absorber. For to assume that only retarded fields, and not advanced fields, can lead to correlated motions among absorber particles would be to commit Price's 'temporal double standard fallacy.'[2] Thus, Zeh cannot allow that there is such a thing as the field physically associated with a charge independently of a specific initial- or final-value problem on pain of begging the question.

At this point of our discussion it is not yet obvious that this is in fact a problem for Zeh's explanation. Yet in the next section I will argue that we can meaningfully ask how changes in a particle's trajectory affect the total field, and that an answer to this question needs to invoke the notion of the field contributed by a charge. A problem for a view like Zeh's is, I will argue, that it can give no answer to the question of how *changes* in the trajectory of charges might affect the total field. Zeh's response must be that the problem is underdetermined without a specification of the new initial or final conditions. "Telling me how the trajectories of a charge change is not enough," he has to say. "You also need to tell me what the new initial conditions for the fields are, and then I will tell you what the field is."

Finally, Zeh's account is problematic in that it proposes to explain initial conditions that appear to be thermodynamically probable by appealing to earlier initial conditions that are absurdly improbable. The explanandum captured by RADASYM is the prevalence of initial conditions with approximately zero incoming fields. Zeh explains this by appealing to the presence of absorbing media that act thermodynamically normally. But the fact that the absorber acts thermodynamically normally in turn needs to be explained by postulating a very low-entropy state in the remote past — that is, by postulating the presence of very large coherent fields in the absorber's past. Thus, Zeh's account explains the weak, fully thermalized fields at the absorber's future boundary, which are thermodynamically very probable, by positing thermodynamically extremely improbable coherent fields in the past. Moreover, Zeh's account does not even allow for these strong coherent fields during the early history of the universe to be explained by the presence of large radiating sources. For without postulating particular initial or final conditions, these fields cannot be associated with any source, on Zeh's view. Intuitively, we would want to say that the large coherent fields resulting in the low-entropy initial state of the absorber are the retarded fields of large radiating sources. Large coherent radiation fields in the early universe, we want to say, are due to the presence of very large and very hot sources.

But this explanation would, of course, require that we can assume that the relevant incoming fields (at even earlier times) are approximately zero. For on Zeh's account, fields can be 'due to' a source only in the context of a given initial- or final-value representation. In an initial-value representation the fields 'due to' the source are retarded, while in a final-value representation the fields 'due to' the source are advanced. And then, of course, the original puzzle arises once more: What can account for the fact that incoming fields at some early moment in the history of the universe were approximately equal to zero? At this point Zeh appears to be faced with a dilemma. He might appeal to his original explanation again, but then a regress threatens. The coherent radiation at past absorber boundaries is explained as the retarded radiation of past sources, which requires the presence of absorbers in *their* past in order to ensure that earlier incoming fields are zero. Yet these past absorbers will in turn have begun their lives in a low-entropy initial state excited by large coherent radiation.

Or Zeh could simply postulate strong coherent fields in the past and leave them unexplained. This second horn of the dilemma, however, is problematic as well. If there were any evidence that the early radiation was in fact retarded, then Zeh's account would not be able to explain this correlation between fields and sources. We do think that there were large sources and that there were large coherent fields in the early stages of the universe. The problem with Zeh's account is that without postulating a specific initial- or final-value problem, there is simply no way to associate the fields with the sources. Zeh has to postulate the large coherent fields and the large sources as uncorrelated and independent. Yet, there is a rival account available that can offer an explanation of the large coherent fields in terms of the sources: an account which assumes that sources are physically associated with retarded fields.

How important it is to Zeh's account to postulate low-entropy initial conditions can also be seen through the following considerations. Imagine a sequence of ideally absorbing regions of space-time alternating with transparent regions, stacked 'on top of each other,' as it were. Then Zeh's account purports to explain why the fields on the future boundary of the *very first* absorbing region are zero, by appealing to a low-entropy past of that absorber and thermodynamic considerations. Now let us postulate that there are no radiating sources present during the first and third transparent periods (following the first and third absorbing periods), but that there is a radiating source during the second transparent period. The transparent periods are separated by ideally absorbing periods. What we would expect to find is that the incoming fields in the second transparent period are zero while the outgoing fields, due to the presence of the source, are not, and are absorbed during the third absorbing period.

How can we account for this asymmetry by appealing to the presence of the absorbers? Notice that both the absorbing region preceding the radiation period and the absorbing region succeeding it are, by hypothesis, in maximum entropy states on their boundaries *not* facing the radiation period. That is, the *second* absorbing period *begins* in a maximum-entropy state, while the *third* absorbing period *ends* in one. The question is What accounts for the fact that the presence of a radiating source leads to a low entropy *past* boundary for the third absorber in the source's future, while the second absorber's *future* boundary in the source's past is unaffected? The point of this thought experiment, of course, is that now an appeal to asymmetric boundary conditions in the second absorber's past are not available, since, to repeat, that absorber's past, like the third absorber's future, is in equilibrium and in a maximum-entropy state. And the worry is that in appealing to statistical considerations here, Zeh would be guilty of a temporal double standard: Without invoking an explicitly time-asymmetric constraint, correlations on the second absorber's future boundary in the presence of a source are no more improbable than correlations on the third absorber's past boundary.

152 Inconsistency, Asymmetry, and Non-Locality

Alternatively, Zeh might argue that our institutions about this case are misguided and that we should not expect radiation to be fully retarded during the second transparent period. But then the worry is that Zeh's account does not have the resources to allow us to determine how the source would couple to the field.

3. The Solution to the Puzzle

The solution to the puzzle of the arrow of radiation that I want to defend can be stated quite simply and consists of the following two theses. First, each charged particle physically contributes a fully retarded component to the total field. This is the *retardation condition*. And second—borrowing a page from Zeh's account—space-time regions in which we are interested generally have media acting as absorbers in their past. This implies that the field on hyperplanes in the not-too-distant future of radiating charges generally is nonzero and, hence, that the field in the past of such hyperplanes cannot be represented as fully advanced. By contrast, the charges involved in a phenomenon do not contribute to the field on hyperplanes in the past of the space-time region associated with the phenomenon. Thus, we can explain why we can generally choose hyperplanes on which incoming fields are zero: The presence of absorbers ensures that radiation fields associated with charges in the remote past do not contribute appreciably to the total field now. That is, fields that are not associated with charges that are relevant to a given phenomenon can generally be ignored, and it is easy to choose initial-value surfaces on which the incoming fields are zero. The asymmetry between initial and final conditions is, thus, explained by appealing to an asymmetric contribution of charged particles to the field and to what, in light of that asymmetry, is thermodynamically to be expected.

The account agrees with Zeh that absorbers play an important explanatory role. Yet, contrary to Zeh, the account does not assume that the absorber condition alone can adequately account for the asymmetry of radiation fields. The brunt of the explanatory work is done by the retardation condition—the assumption that the field physically contributed by a charge is fully retarded. Moreover, unlike Zeh's account, the explanation need not appeal to the statistically improbable. While it is possible for absorbers to act anti-thermodynamically, zero fields at a future boundary of an absorbing region of space-time are what is thermodynamically to be expected, given the retardation condition. The fact that charged particles contribute retarded rather than advanced fields explains why absorbing media act as absorbers rather than as anti-absorbers. Retarded fields associated with charges are responsible for any coherent radiation at an absorbing region's past boundary, while there is no mechanism that can make coherent excitations at the absorber's future boundary probable. Thus, even though the dynamical laws governing the absorber are time-symmetric, initial and final conditions at the absorber boundaries are not the temporal inverses of one another, and this can explain the temporal asymmetry of absorbers.

Why should we accept this explanation of the asymmetry of radiation? Many physicists and philosophers writing on the subject of the arrow of radiation are dissatisfied with an explanation along the lines I proposed, and find it mysterious.

The following appear to be the main worries concerning this explanation. First, simply postulating the retardation condition as additional constraint strikes many as not very illuminating. While appealing to the condition may help explain why incoming, but not outgoing, fields can generally be set to zero, this explanation immediately seems to raise yet another puzzle: Why, given that the Maxwell equations are time-symmetric, does the retardation condition *itself* hold? Where does this additional asymmetric constraint come from? Since it is not implied by the putative laws of the theory, the constraint itself, it seems, is in urgent need of an explanation if we are to find this explanation of the asymmetry of the total fields satisfactory. Yet this further puzzle remains unanswered by the account I am advocating. Elsewhere I have offered some reasons why we ought to be suspicious about the way in which this demand for an explanation is generated and have argued that the mere fact that there may be a constraint on electromagnetic systems not implied by the Maxwell equations is not itself sufficient to generate the need for an explanation (see Frisch 2000).

Second, one might worry that there seems to be 'no room' in the theory for any constraints in addition to the Maxwell equations. The Maxwell equations alone define a complete initial- (or final-) value problem and there appears to be no need to appeal to any additional constraints. We should not expect more from a theory, one might say, than that it provides us with a well-defined initial-value problem that determines how states of the world evolve in time. And third, the notion of a field physically associated with a charge might appear illegitimate since it implies counterfactual claims that might strike one as problematic. In the next section I will spell out the notion in more detail and will try to dispel these last two worries.

4. The Retardation Condition as Causal Constraint

What do I mean by the notion of the field physically associated with a charge? As a preliminary characterization I want to offer the following: The field component associated with a source is that component of the total field which would be absent if the source were absent. What that component is, is independent of any particular initial- or final-value problem. Imagine a region of space-time (that is not bounded by an absorber in the past) with no charges but with an arbitrary nonzero electromagnetic free field. If the region is source-free, then the field F_{in} on an incoming hypersurface will be related to the field F_{out} on an outgoing surface via the source-free Maxwell equations. We can then ask how the total electromagnetic field would change if a charged particle were introduced into the region. From a physical standpoint the situation appears to be completely determined: Initially the total field is a known source-free field. Then a charge (with a trajectory that we assume to be known) is introduced. The resulting total field should be given by the sum of the source-free external field and the field associated with the charge and its motion. And the retardation condition tells us that the field physically associated with a charge is fully retarded, rather than being some linear combination of retarded and advanced fields.[3]

I take it that the notion of a retarded field physically associated with a charge to which I am appealing is essentially the same notion as the one invoked by Einstein when he speaks of "elementary radiation processes" (Einstein 1909b). According to Einstein, whose view we examined in chapter 5, elementary classical radiation processes are time-asymmetric and consist of diverging waves. What each elementary 'generator' of a wave—that is, each accelerating charged particle—contributes to the total field is a diverging wavelet. Of course diverging wavelets can mathematically also be represented as sums of source-free fields and converging wavelets. Yet the question of what the field associated with an elementary radiation process is, can be addressed completely independently of any specific initial- or final-value problem. As we have seen, the idea that each charge physically contributes a certain field (or is physically responsible for a certain force on all other charges) is also embraced by advocates of absorber theories of radiation in the tradition of Wheeler and Feynman, who do not, of course, assume a retarded field but take the 'field' associated with a charge to be symmetric, half retarded and half advanced. In fact, implicit appeals to this notion are pervasive in the literature on radiation, and even those who explicitly disavow the notion of a field physically associated with a charge sometimes end up making use of it.

The Maxwell equations have both retarded and advanced solutions, but according to the retardation condition, only the former represents the physical contribution of a charge correctly. Thus, according to the account I am defending here, the retardation condition provides an important constraint on electromagnetic fields in addition to those embodied in the Maxwell equations. Now, as I said above, there are two worries one might have about this additional constraint. First, one might doubt whether there is in fact room in the theory for any such additional constraint. Since the Maxwell equations pose a well-defined initial-value problem, they completely determine how the state of the electromagnetic field evolves, given the fields on an initial-value surface together with the trajectories of all charged particles. That is, once the initial or final values and the particle trajectories are given, nothing but the Maxwell equations is needed to determine how the electromagnetic field evolves forward or backward in time. There is, one might argue, nothing more we should expect from a deterministic theory than a unique answer to the question of how the state of the world at one time determines the state of the world at all other times. Now, in fact, in addition to providing diachronous constraints, the Maxwell equations also constrain possible instantaneous states of the world—the magnetic field has to be divergence-free, while the divergence of the electric field is given by Coulomb's laws. But important for our purposes here is that the diachronous constraints tell us that if the field is given in terms of an *initial*-value problem, then charged particles contribute *diverging* waves, and that if the field is specified in terms of a *final*-value problem, charged particles contribute *converging* waves. Since these waves, in conjunction with the free fields on the initial-value surface, determine the state of the field completely, there might appear to be no place in the theory for an additional notion of the 'physical contribution' by a charge to the field along the lines I suggested.

The second worry is this. My initial gloss of the notion of the field physically associated with a charge was in terms of counterfactuals: The field physically associated with a charge is that component of the field which would be absent if the charge were absent. But counterfactuals are often regarded with suspicion and are thought to be hopelessly vague and without clear meaning. So one might worry that even if the retardation condition could be tacked onto the theory, that condition, as causal or counterfactual condition, is too ill-defined to be acceptable as a legitimate ingredient of a scientific theory.

This last objection concerning the status of counterfactuals could be motivated by several different concerns. One may have very general worries about the status of counterfactuals and may find that any appeal to counterfactuals is illegitimate in the context of fundamental science. Yet every theory with laws that are differential equations defining a pure initial-value problem makes counterfactual claims about the system governed by its laws: The theory not only determines the actual evolution of a system but also provides an answer to the question of how the evolution of a system would change if the initial (or final) values characterizing the system changed. That is, in the case of a pure initial-value problem, we can ask how changes in the initial values would affect the system, and the theory provides a well-defined answer to this question. A common complaint against counterfactuals is that they are hopelessly vague and context-dependent. But the counterfactuals implied by a theory's equations do not seem to face this problem, since the theory gives a precise and unique answer to the question of how changes in initial conditions affect the future or past evolutions of a system. Thus, counterfactual claims are part of the content of scientific theories (even though we might want to have a different *attitude* toward the counterfactual claims than toward the theory's predictions concerning actual systems) and, hence, any qualms about the retardation condition cannot be merely general worries about the status and legitimacy of counterfactuals. If the notion of the field physically associated with a charge is problematic, since it supports certain counterfactuals, these problems have to be related to the nature of the specific counterfactual claims involved in this case.

A more specific concern is that in assessing how the total field would be affected by the presence of an additional charge, we are asked to consider the effects of a counterfactually 'created' charge, and that this questions does not seem to permit well-defined answers in the context of classical electrodynamics. Since the Maxwell equations imply that charge is conserved, postulating the creation of an additional charge seems to take us beyond the confines of the theory. This worry, however, can easily be met as well. First, we have seen in chapters 2 and 3 that a theory can generate well-defined predictions even from incompatible assumptions. Second, in assessing the effect of an additional charge, we can evaluate the fields in a world where the additional charge always was present—that is, a world in which charge is conserved. Finally, and most important, instead of considering the effects of additional charges on the field, we can motivate the notion of the field associated with a charge by asking how the field in a region containing a charge moving on a certain trajectory would change if the trajectory were altered, perhaps due to nonelectromagnetic forces. Addressing this question does not require us to contemplate electromagnetic 'miracles.' The change in field strengths is given by the difference in the fields associated with the charge on its different trajectories. Thus, the fact that the retardation condition implies certain counterfactual claims does not seem to be any more problematic than the fact that scientific theories in general involve such claims.

What about the worry that there is no room in the theory for constraints in addition to the Maxwell equations? Not only is there room for the retardation condition as an additional constraint on the theory, but the theory's content would in fact be unduly restricted without that condition, for the condition allows us to address certain scientifically legitimate questions that in a theory consisting of the Maxwell equations alone must remain without answer. In general, there are two contributions to the total field at a given space-time point: source-free fields and fields associated with charged particles. The two contributions are modeled in classical electrodynamics as being independent and independently alterable. Thus, in evaluating the effects of the source-free fields, we can keep the trajectories of the charges fixed and ask how changes in the initial field will affect the total field. This type of question can be addressed with the help of the Maxwell equations alone, by solving a new initial-value problem for the system of fields and charges. But we can also ask how changes in the charge distribution would affect the total field. If the former kind of question is legitimate, so should the latter be, since both questions concern the effects of changes in the input into the theory's system of equations. The Maxwell equations *alone*, however, do not provide an answer to the second type of question, since in order to be able to use the equations, we first have to know what effect such changes have on the initial or final conditions. Here the retardation condition plays an important role: It tells us that changes in the trajectory of a charge will have no effect on initial conditions, but only on final conditions, and that the difference in the final fields is given by the difference in the contributions of the *retarded* fields associated with the different trajectories of the charge. By contrast, on a conception of classical electrodynamics without the retardation condition (or without any other condition fixing the field physically associated with a charge), there is no general answer to the question of whether the acceleration of a charge affects the fields on the future light cone of the charge or its past light cone.

Consider the problem of determining the total field in the presence of a charge that is experiencing a brief (nonelectromagnetic) force in a known nonzero incoming free field. The total fields can, of course, be determined by setting up an initial-value problem and using the Maxwell equations. What, however, would the electromagnetic field be if the force pulse had been twice as large? Without recourse to the retardation condition, the answer has to be "We simply cannot know, since the theory provides no answer to this question. Before we can determine the fields, we first need to be given the new initial or new final conditions." That is, if we are not able to invoke the notion of the field physically associated with a charge, the problem of how changes in the charge distribution affect the total field is ill-defined. By contrast, a theory that includes the retardation condition provides us with equally well-defined answers to the question of how changes in initial fields affect the total field *and* to how changes in the charge distribution affect the total field. Thus, I agree with Hogarth's claim that a theory that does not specify the field associated with a single source is incomplete (see Hogarth 1962, 366).

The field *physically* associated with a charge need not coincide with the field *mathematically* associated with a charge in a particular representation of the total field. Consider an initial-value problem with zero incoming fields—that is, a situation with a fully retarded field. This field can alternatively be represented in terms of a final-value problem as the sum of an advanced and a source-free outgoing field. One might ask where the source-free outgoing field can come from, if there is no source-free incoming field. The answer is that the charge (or charges) of the problem are responsible for that field. For a space-time point on the worldline of a charge at a time t_0, the associated retarded field is zero for times $t < t_0$, while the advanced field is zero for times $t > t_0$. Thus, if a field equal to a fully retarded field is to be represented in terms of advanced fields, this representation has to include an outgoing field which represents the field after the advanced field has 'turned off.' But if we take the fully retarded field to be (physically) due to the charges and their motions, then one should likewise take the partly advanced, partly free field to be due to the charges. After all, the field is one and the same, and only its mathematical representation has changed. Thus, a particular representation of a field might be somewhat misleading, in that even components of a field that do not have a source in a certain space-time region can be physically associated with a source in that region.

There is a certain danger in reading claims about the independent existence of fields into a particular mathematical representation of these fields. It does not follow from the fact that we can rewrite a fully retarded field associated with a source as the sum of an advanced field and a source-free field, that there is a source-free field which exists independently of the source. If the source were absent, so would be the field. There is of course not an identifiable part of the total field strength that belongs to a given charge. We cannot say that, for example, the first two-thirds of the field vector at a certain point belong to a charge, while the rest is part of a free field. But this does not mean that we cannot take each charge as contributing a definite amount to the total field, where that amount is, on the view I am defending here, given by the retarded field associated with each charge.

I have introduced the notion of the field physically associated with a charge by appealing to the counterfactuals supported by that notion. Yet I believe that fundamentally we should view the retardation condition as a *causal constraint*, according to which accelerating charged particles are causes of disturbances in the electromagnetic field which propagate along future light cones. Recall Einstein's principle of *local action* that we discussed in chapter 4. Einstein's principle concerns the question of how *external influences* on one part of a system propagate through the system; if a theory satisfies the principle of local action, then, according to the theory, for two spatially distant things A and B, "externally influencing A has no *immediate* influence on B" (Einstein 1948, 321). As we have seen, this principle does not reduce to a pure principle of determinism, such as Schrödinger's 'principle of causality.' And perhaps this should have been obvious from the beginning, since clearly a local substate of a system at one time can determine a substate of the system at some other time without it being possible to influence the latter state through manipulating the former from the outside.

Analogous to Einstein's causal locality principle, the causal character of the retardation condition manifests itself in how a particle–field system may be influenced from the outside. According to the retardation condition, an external influence on the state of motion of a charge affects fields in the future but not in the past. Similarly, there is a causal link between initial and final fields: Externally influencing the initial state of the field has an effect on later states of the field, but external influences of the final state do not affect earlier states. Thus there is a causal asymmetry between initial and final conditions. This asymmetry implies, further, that we can set up the initial fields independently of any concern for, or later interventions into, the particle trajectories, but we cannot similarly set up final fields independently.

One might object to this and maintain that there is no such asymmetry. While it is true that by *holding the initial conditions fixed*, the final fields vary with and are determined by the trajectories of the charged particles, it is equally the case that by *holding the final conditions fixed*, the initial fields vary with and are determined by the trajectories of the charged particles. Or, putting the point slightly differently, for each system and each counterfactual intervention into its particle trajectories, there is one counterfactual system that has the same initial fields but different final fields *and* one counterfactual system that has the same final fields and different initial fields in comparison with the original system. And indeed, one might argue that this further shows that counterfactuals associated with changes in particle trajectories are ill-defined after all: There is no unique, context-independent answer to the question of whether changes in a particle's trajectory affect future fields or past fields. A well-defined context is given only by further specifying whether one wants to hold initial or final conditions fixed.

But, of course, once we introduce the retardation condition, there no longer is an asymmetry between initial and final conditions, and the vagueness disappears: The retardation condition tells us that it is the initial conditions, and not the final conditions, that remain unaffected by changes in the particle trajectory. Still, one might ask, what justifies our adoption of the retardation condition? It seems to me that, similar to the case of the Maxwell equations, our experimental interactions with electromagnetic systems support the claim that initial conditions are causally independent from the later trajectories of charges, while final conditions are not. 'Counterfactual' systems with initial conditions identical to their actual values but different particle trajectories are, as it were, easily accessible from the actual world; we merely have to conduct a second run of an experiment with identical initial conditions for the fields. Yet counterfactual systems with final conditions identical to those of the actual system are much harder to reach. That is, our experimental interactions with systems of charged particles suggest that we can set up initial conditions independently of the acceleration of later charges and that we can alter the acceleration of a charge without that having an effect on the initial conditions. And we can then use these initial conditions in conjunction with knowledge of the trajectory of the charge to determine the fields everywhere.

By contrast, keeping the final conditions constant in spite of changes in the acceleration of a charge would require a careful balancing act of introducing different compensatory effects. For example, we can imagine that two experimenters are responsible for setting up initial fields and particle trajectories, respectively, and that each does so without knowledge of what states the other chooses. But it is impossible for an experimenter to set up a specific final field without taking into account the contributions of charged particles in the past. That is, any experimenter responsible for ensuring that final fields have a certain value needs to know what other experimenters chose for the particle trajectories in the past of the final-value surface. And if we had some means of setting up arbitrary final conditions 'by brute force,' independently of knowledge of the prior fields (for example, by means of an absorber that absorbs all incoming fields), these final conditions could no longer be used to calculate the fields in their past. This experimental asymmetry is readily explained by the claim that each charged particle contributes a fully retarded field. The retardation condition, thus, provides an answer to Zeh's question of why initial conditions are more useful than final conditions (see chapter 5): Since the field contributed by a charge is fully retarded, we can control initial conditions independently of the motion of a charge in ways in which we do not have independent control over final conditions.

That incoming fields and the charge distribution are treated as independently alterable causes of the total field is strongly suggested by the way radiation phenomena are modeled. Instead of determining the state of an electromagnetic system in terms of a pure initial- (or final-) value problem, radiation phenomena are modeled as retarded field problems for which the full trajectories of all charged particles in the space-time region at issue need to be given in addition to the fields on an initial-value surface. Thus, models of radiation phenomena do not satisfy John Earman's condition for Laplacian determinism. Recall that according to Earman's condition, worlds are deterministic just in case if any two worlds agree on one Cauchy surface, they agree everywhere (Earman 1986, 59).[4] Since the trajectories of a charge are treated as independent input, and not as being determined through an initial-value problem in models of radiation phenomena, there can be two models that agree on one Cauchy surface but disagree on other surfaces, due to a disagreement in the trajectory of the charge. Nevertheless, the theory is deterministic in a different sense: If in addition to the fields on one time slice *the full trajectories of all charged particles agree* in two models, then the fields in the two models agree everywhere.

One might think that the reasons for representing the field in terms of retarded (or advanced) fields and fixing the trajectories independently of (and prior to) determining the field are entirely pragmatic. For one, it simplifies the mathematics: The Maxwell equations are linear, while possible candidates for complete particle–field equations are not. Moreover, a charged particle could in general also be subject to nonelectromagnetic forces. This means that in order to specify a full initial-value problem, we would have to include a theory governing the temporal evolution of any relevant nonelectromagnetic system coupling to the particles as well. But the details of such a theory would be irrelevant to questions concerning the properties of radiation fields. Thus, modeling an electromagnetic phenomenon in terms of a retarded- or advanced-field problem has the advantage of enabling us to isolate the purely electromagnetic aspects of the phenomenon and, in particular, allows us to focus on how the motion of charged particles affects the state of the electromagnetic field.

But there is another reason for treating the trajectories as independent that prevents pragmatic considerations, important as they may be in general, from becoming relevant in this case: There is no well-defined and conceptually unproblematic way of setting up a pure initial-value problem for charged particles in an electromagnetic field, since this would require taking into account the effect of the total field on charged particles as well. To set up a pure initial-value problem, we would need an equation of motion for a charged particle in an electromagnetic field. As we have discussed in some detail in chapters 2 and 3, the two most plausible choices for an equation of motion are the Lorentz equation of motion, which is a Newtonian force law, and the Lorentz–Dirac equation. Neither of the two equations is unproblematic, however. Adopting the former results in an inconsistent theory, while the latter is conceptually deeply problematic and there are no general existence and uniqueness proofs for solutions to the latter for systems consisting of more than two particles. There is a well-posed initial-value problem for *continuous* charge distributions, yet even this case is problematic, as we have seen, since for many intuitively reasonable initial conditions the equations do not admit of global solutions. Moreover, there does not appear to be any complete and consistent way of modeling the behavior of discrete localizations of charges in the continuum theory.[5]

There is then, I take it, no plausible alternative to the standard approach of modeling electromagnetic phenomena by treating the two types of interaction between charged particles and electromagnetic fields—forces acting on a charge and electromagnetic disturbances due to charges—separately. In particular, if we are interested in properties of the radiation field, the trajectories of charges are assumed to be fixed. Within that approach it is irrelevant to the question of how charged particles affect the electromagnetic field whether or not these trajectories could themselves be given in terms of an initial- or final-value problem. As I have suggested in chapter 2, this practice of modeling electromagnetic phenomena is naturally understood as treating charged particles and fields as two subsystems of electromagnetic systems that interact with one another through two causal routes: Fields have an effect on the motion of charged particles, as given by the Newtonian Lorentz force equation of motion, and the motion of charged particles in turn has an effect on the total electromagnetic field, as given by the Maxwell equations in conjunction with the retardation condition.

For modeling purposes these two routes of causal interaction are treated separately. We can investigate the effects that charged particles have on the total field, while ignoring possible effects that the field will have on the motions of the particles. I take it that this selective treatment is much harder to understand on an interpretation of the theory that tries to eschew all causal talk. If all that the equations present is one way of capturing the functional dependencies governing interacting charges and fields, it becomes unclear why it is precisely the Maxwell equations and the Lorentz law that can be used in isolation to describe phenomena involving charged particles. On the causal interpretation I am defending here, there is no such difficulty: The two equations govern different causal processes that can be modeled in isolation from one another.

5. Causes and Interventions

In the last section I proposed that the retardation condition is a causal constraint, and I suggested that such a causal constraint is supported by our experimental interactions with electromagnetic systems. In fact, the role of the retardation condition in modeling radiation phenomena that I outlined fits rather well with interventionist or manipulability accounts of causation (see, e.g., Hausman 1998; Pearl 2000; Woodward 2003). I want to end this chapter by briefly summarizing some of the salient features of such accounts and showing how they apply to the interactions between classical particles and fields.

Consider a system that can be completely closed or can interact in a limited, isolated manner with its environment. We assume that when the system is closed, the laws governing the system determine completely how the state of the system at any one time depends on the state of the system at other times. Within the closed system, various relations of determination will hold, and considering *only* the completely closed system does not enable us to attribute causal relations to the system going beyond claims of determination. Relations of causal influence are revealed by comparing the closed system with the ways in which copies of the system are affected by isolated interactions with the system's environment. Such limited interactions have the effect that certain variables characterizing the system take on values other than they would have taken on if the system had been completely closed.

In the literature on causal modeling, variables or changes in the values of variables are usually taken to be the relata of the causal relation. Applied to the case of radiation, the dynamical variables characterizing the state of a charged particle—that is, its position, velocity, and acceleration—are causes of **E** and **B**, the variables characterizing the state of electromagnetic field and changes in the particle variables cause changes in the appropriate field variables. The basic idea of interventionist accounts of causation is that the claim that a variable X is a cause of a variable Y ought to be spelled out in terms of the effects of possible interventions that change the value of the variable X. Roughly, X is a cause of Y if and only if possible interventions that cause changes in the value of X result in changes to the value of Y. This is only a rough consideration, for the condition needs to be supplemented by constraints which ensure that the intervention on X affects Y along the correct causal route. For example, we need to exclude interventions that causes a change in the values of both X and Y but act as common causes of X and Y. For our purposes, however the preliminary characterization will be sufficient.[6] Note, also, that interventionist accounts are not meant as a reduction of the notion of causation, since the conditions for an intervention themselves appeal to causal relations. Rather, the hope in appealing to the notion of an intervention is that exploring the connections between this concept and that of cause (and perhaps other related concepts) can help to illuminate the notion of cause.

On an interventionist account, experimental interventions into a system are an important guide to the causal relations instantiated in the system. In fact, one might think of interventions as the effects of possible idealized experiments by which one can establish the causal relationship between two variables X and Y. Let us assume that the values of the variables X and Y vary with one another in systems of a certain type, and we are interested in finding out whether the correlations between the values of X and Y are the result of a causal relationship. According to an interventionist account, we can determine what (if any) the causal relationship between X and Y is by intervening on X. If the value of Y can be manipulated through causing changes in the value of X, then X is a cause of Y. Yet while thinking about idealized experiments is a useful tool, the notion of *human* interventions is not, however, essential to the account. What is essential is that the system whose state is partially characterized by the values of the variables X and Y, is isolated or closed *except for the intervention* on the variable X. An intervention can be any isolated causal interaction between the system and its environment that affects the value of X. Through the intervention the link between X and what otherwise would have been its causal history is at least partially broken. That is, if we characterize interventions in terms of the effects of an intervention variable *I* (characterizing the state of the system's environment), then X has as a cause — either a complete cause or a contributing cause — the intervention variable *I* such that the value of X is no longer determined exclusively by the values of whatever other variables characterize its causal past within the system.

On an interventionist conception of causation, causal relations are assessed by looking at situations where the variable intervened upon takes on values that are different from the value it would have in a closed, deterministic system. That is, the value of the variable intervened upon is allowed to vary independently of the causal history of the system intervened upon. This conception can quite naturally account for the way in which retarded field problems are modeled in electrodynamics, where particle variables are treated as independently alterable and not determined through an initial-value problem. Particle–field systems are treated as closed systems in which the incoming (or outgoing) field and the particle trajectories determine the field everywhere, while the particle trajectories themselves are treated as independently given. For the purposes of modeling radiation phenomena, we can set the variables characterizing the state of a particle to any values we like without having to take into account other variables that might (in a different type of model) affect the state of motion of the particle. Hence the models naturally suggest an interpretation in terms of possible interventions on the values of the particle variables: Since the entire trajectories are treated as input that is not determined by the state of the field, we can 'intervene' anywhere to change values of the variables characterizing the trajectories. One justification within electrodynamics for this modeling practice is that while the field in principle has an effect on the motion of the charged particles (even though there is no consistent way of modeling this), the particles can be affected by other, nonelectromagnetic forces as well, which can result in interventions on the state of a charge.

How, then, do interventions on the particle variables affect the state of the field? As we have seen, the notion of the field physically associated with a charge is needed to determine how interventions on the particle trajectories percolate through the system of charges and fields. In particular, the retardation condition specifies that the causal route by which particle variables affect the value of field variables is along forward light cones. Thus, models that contain a subset of variables whose values through time are independently alterable (and are not fixed through an initial-value problem), such as the particle variables in radiation models, strongly suggest a causal interpretation. And contrary to the suggestion one sometimes finds in the literature that thinking of causation in terms of interventions is appropriate only in the context of applied science but not in fundamental physics, modeling practices at the 'highest' theoretical level in classical electrodynamics support a manipulability view of causation.

The other type of models in classical electrodynamics—models that use the Lorentz force equation of motion to represent the effects that fields have on charged particles—fits intervention accounts equally well. These models focus on the causal route from field forces to motions of charged particles and correspondingly allow the forces acting on the charge to be independently alterable. Particle motions are determined from initial values of the particle variables in conjunction with the forces the particles experience anywhere. As in the case of the particle variables in radiation models, field variables are treated as independent inputs in Lorentz force models.

Advocates of interventionist accounts of causation often seem to believe that their account points to certain limits to positing causal relations. Since the notion of intervention crucially relies on distinguishing a system from its environment, one might think that there is room for causal notions only in the 'special sciences' that do not purport to present us with models of the world as a whole. Fundamental physics, on the other hand, does, it seems, aim to provide us with 'global' models of possible ways the world could be; and because global models have no environment or 'outside,' an interventionist notion of cause cannot apply here. I want to end this section with two brief remarks about this line of reasoning.

First, as we have seen in the first part of this book, despite what initial appearances may suggest, even 'fundamental' physical theories need not provide us with models of possible worlds as rich as ours. Classical electrodynamics allows us to construct models of particular phenomena, yet these models cannot be integrated into 'global' models of electromagnetically possible worlds. Thus, the antecedent of the limiting argument is not satisfied. Modeling in fundamental physics may be a lot more like modeling in the special sciences than this line of thought suggests. Second, even if we had a fundamental physics with global models, this does not yet make it impossible for causal notions to be applicable, as long as the theory also has models of finite systems. For causal relations could then be supported by experimental interactions with such systems. Just as we confirm the laws we take to govern the universe as a whole through interactions with finite subsystems, so we may be able to confirm causal relations holding in the world through interactions with such subsystems.

6. Conclusion

In chapter 5 I argued that the asymmetry of radiation phenomena is an asymmetry between initial and final conditions. Physics textbooks usually appeal to causal considerations to account for this asymmetry. Most of the discussions of the asymmetry in both the physics and the philosophy literatures, however, reject such an account. The widespread dismissal of a causal account appears to be due to intuitions similar to those discussed in chapter 1, according to which the content of scientific theories is limited to what is implied by a theory's fundamental equations. If, as in the case of the radiation phenomena, not all of a theory's state-space models represent phenomena which we might possibly observe—that is, if there is a large class of models that are compatible with the theory's basic equations yet do not represent actual phenomena—then this has to be explained through some other 'legitimate' physical mechanism, according to the standard view, and could not simply be due to a general asymmetric causal constraint.

In chapter 6 and at the beginning of this chapter I examined what seem to be the main proposals in the literature that try to account for the radiation asymmetry in terms of some other physical asymmetry. As we have seen, none of these proposals is ultimately successful. I then argued that the asymmetry between initial and final conditions characterizing radiation phenomena can best be explained by appealing to a fundamental causal constraint: The field physically associated with an individual charge is fully retarded. If in addition we assume the presence of media which can act as absorbers, we can explain why it is often possible to represent radiation phenomena involving a relatively small number of sources in terms of an initial-value problem with zero incoming fields, while only rarely can we find representations involving only relatively few charges where outgoing fields are zero.

One argument for the causal account I am advocating, then, is negative: There is no rival account available that is equally successful. But I also offered some positive reasons appealing to our experimental interactions with electromagnetic systems for accepting the account. We appear to have good evidence that interventions on particle trajectories affect the field along future light cones and not along past light cones. Any account of the asymmetry ought to be able to account for this fact.

If it is indeed correct that the best explanation of the asymmetry of radiation appeals to a causal constraint, then this further supports my claim that standard accounts of theories which strictly identify theories with a set of dynamical laws and a mapping function specifying the theory's ontology are too thin. Contrary to what Russell has famously argued, 'weighty' causal claims seem to play an important role even in fundamental physics.

8

David Lewis on Waves and Counterfactuals

1. Introduction

In the previous three chapters I examined possible answers to the following question: Why electromagnetic fields in the presence of charged particles can generally be represented as fully retarded but not as fully advanced—that is, as coherently diverging but not as coherently converging waves? I argued that the best explanation for the fact that it is generally possible to assume that incoming fields (but not outgoing fields) are equal to zero is that in addition to the Maxwell equations, there is a causal constraint on electromagnetic phenomena—*the retardation condition*, according to which each accelerated charge contributes a diverging wave to the electromagnetic field. In this chapter I want to try to correct a prominent misunderstanding of the wave asymmetry. David Lewis has cited the asymmetry of wave phenomena as the central example of his thesis of the asymmetry of overdetermination, according to which events leave many traces in the future that individually postdetermine the event, but there are not similarly multiple 'traces' in the past that individually predetermine an event's occurrence. This thesis, I want to show, is false.

The failure of Lewis's thesis of overdetermination has broad repercussions for his accounts of causation and of counterfactuals. Thus, this chapter also continues the exploration of certain themes from chapter 4: Lewis's counterfactual theory of causation is an attempt to provide a broadly neo-Humean account of causation that is meant to avoid the difficulties faced by regularity accounts. Lewis's theory is a Humean theory not only in that it arguably takes one of Hume's definitions of cause as its cue, but also, and more importantly, in that, according to Lewis, it supports the thesis of *Humean supervenience*, according to which "all there is to the world is a vast mosaic of local matters of particular fact" (Lewis 1986c, ix) that do not include irreducibly causal facts. Thus, someone might want to appeal to Lewis's theory in response to my discussion in chapter 4 in support of some type of Humean or traditionally empiricist account of causation. A Russellian notion of causation as functional dependency may not be rich enough to account for the role of causal reasoning in science, as I argued in that chapter. But perhaps an only slightly richer account, such as Lewis's, may be adequate. In trying to motivate a particular causal interpretation of Dirac's classical theory of the electron (in chapter 4) and in arguing for the need to posit the retardation condition (in chapter 7), I myself have appealed to counterfactual reasoning. If a counterfactual *analysis* of causation were successful, then there would be no need to posit irreducibly causal relations, even if scientific theorizing appeals to notions of cause that cannot be reduced to the notion of functional dependency.

I will argue, however, that Lewis's analysis cannot adequately account for the asymmetry of causation. In the next section I will present a brief overview of Lewis's account that focuses on the roles of the theses of an asymmetry of miracles and of overdetermination in explaining the asymmetry of causation. Then I will discuss an earlier criticism of Lewis's overdetermination thesis due to Frank Arntzenius. Arntzenius's criticism points to ways in which Lewis's account has to be sharpened but does not offer a knockdown argument against the account.

In section 4, I will offer my own criticism of the overdetermination thesis, arguing that the asymmetry of wave phenomena supports neither an asymmetry of overdetermination nor an asymmetry of miracles. Since Lewis's account of an asymmetry of counterfactuals relies on these asymmetries, the account is unsuccessful as it stands. Nevertheless, there might be a way to rescue a possible worlds account of counterfactuals on which backtracking counterfactuals come out false, if Lewis's criteria by which we judge the similarity of possible worlds are amended to include agreement in broad qualitative features, such as those embodied in phenomenological thermodynamic 'laws.' For, as we will see, worlds perfectly converging to worlds like ours are thermodynamically peculiar. Yet, since this qualitative difference between diverging and converging worlds is due to the fact that the retardation condition holds at these worlds, and since this condition ought to be understood as a causal constraint, my friendly amendment to Lewis's account of counterfactuals is in fact a 'poison pill' for his account of causation.

In section 5, I criticize Daniel Hausman's argument for an overdetermination thesis that is closely related to Lewis's. In section 6, I will briefly discuss a criticism of Lewis's account due to Adam Elga, whose discussion of the asymmetry of miracles in the context of the foundations of thermodynamics further supports my conclusions. I end with some concluding remarks.

2. Causation, Miracles, and Overdetermination according to Lewis

The fact that the causal relation is asymmetric is often taken to pose a serious problem for regularity theories of causation, which might try to solve the problem by simply adding the condition that causes precede their effects to an otherwise symmetrical analysis of the causal relation. This was Hume's strategy. But if we want to allow for the conceptual possibility of backward-causal theories, such as Dirac's classical theory of the electron, which we discussed in chapters 3 and 4, this strategy is inadequate. What we need is an account that does not settle the temporal direction of causation by definition.

Lewis's account of causation appears to be superior in this respect. To summarize briefly, Lewis introduces a relation of causal dependence that he defines in terms of counterfactual dependence: For two actual events C and E, E *causally depends* on C, just in case that if C had not occurred, E would not have occurred. The relation of causation is the ancestral relation of that of causal dependence: C *is a cause of* E just in case there is a chain of causally dependent events that link C and E (Lewis 1986a). Lewis does not directly define the relation of causation in terms of counterfactual dependence in order to be able to account for cases of overdetermination, where a second possible cause is 'waiting in the wings' to bring about E in the absence of C. As is well known, Lewis's account needs more 'tweaking' to account successfully for various possible cases of overdetermination, preemption, or 'trumping'; and these types of problems for Lewis's original analysis have led him to revise his account substantially.[1] But these issues need not concern us here, since they do not affect Lewis's explanation of the asymmetry of causation.

The asymmetry between cause and effect, according to Lewis, is due to the fact that the relation of counterfactual dependence is asymmetric. On Lewis's account, the truth conditions of counterfactuals are given in terms of possible worlds and a similarity metric is defined on the class of possible worlds: The counterfactual 'If C had not occurred, E would not have occurred' is nonvacuously true, exactly if there is some non-C world in which E does not occur that is closer to the actual world than any non-C world in which E does occur. The closeness between worlds is measured by some weighted average of both agreement in laws and agreement in particular matter of fact. In assessing the similarity of different possible worlds, according to Lewis,

(1) It is of the first importance to avoid big, widespread, diverse violations of law.
(2) It is of the second importance to maximize the spatio-temporal region throughout which perfect match of particular fact prevails. (Lewis 1986b, 47–48)[2]

Thus, for Lewis the non-C world closest to the actual world is not one in which the laws of the actual world hold completely (and which therefore has both a past and a future different from the actual world, if the laws are deterministic), but rather one that matches the actual world right up to the occurrence of C, but from which C is then removed by a small 'miracle,' after which the laws of the actual world are once again allowed to run their course.

There is an asymmetry of counterfactuals, since backtracking counterfactuals—such as 'If E had not occurred, C would not have occurred,' where C precedes E—generally and in standard contexts come out false on Lewis's account. For Lewis maintains that there is no non-E world in which C does not occur that is closer to the actual world than all non-E worlds in which C does occur. There are two kinds of worlds in which neither E nor C occurs which we need to consider. One kind is worlds that have the same past as the actual world but in which a miracle prevents C (and hence E) from occurring. Such worlds diverge from the actual world *before* the occurrence of C. Lewis argues that it would be less of a departure from actuality to wait until *after* the occurrence of C, and then introduce a miracle that prevents E from occurring, than to introduce an earlier miracle that prevented C from occurring. For close C worlds without E still contain many of the other future traces C leaves in the actual world, while the closest non-C worlds do not. Thus there is a closer match between close C worlds and the actual world than between close non-C worlds and the actual world. The later divergence occurs, the closer the match between the actual world and the diverging world.

The second kind of world to consider is worlds in which neither C nor E occurs and which have histories different from the actual world, but which converge with the actual world right after E occurred in the latter. The future in a converging world is exactly like the future in the actual world, while the past is not. Lewis maintains that converging worlds will generally be less similar to the actual world than diverging worlds are. The reason for this, according to him, is that divergence from the actual world is much easier to achieve than perfect convergence to it. For a small, localized miracle is enough to result in a world that diverges from ours. But, Lewis believes, to achieve perfect convergence, a world with a past different from that of the actual world would need somehow to produce the multitude of traces that past events leave in the actual world. And this would require a big miracle. Thus, the counterfactual asymmetry is due to an *asymmetry of miracles*. A tiny miracle is enough to lead to divergent worlds, but perfect convergence requires a "widespread and complicated and diverse" miracle (Lewis 1986b, 46).

The asymmetry of miracles, finally, is a consequence of an *asymmetry of overdetermination*, according to Lewis. He maintains that, at least in worlds as complex as ours, earlier affairs are extremely overdetermined by later ones. Lewis believes that "whatever goes on leaves widespread and varied traces at future times" (Lewis 1986b, 50), and that even small subsets of these traces could not have come about in any other way than through the event of which they are in fact traces. Thus, "it is plausible that very many simultaneous disjoint combinations of traces of any present fact are determinants thereof; there is no lawful way for the combination to have come about in the absence of the fact" (Lewis 1986b, 50).

Here a *determinant* of a fact is a minimal set of conditions, holding at a particular time, that is jointly sufficient, given the laws at the world in question, for the fact in question. The *thesis of overdetermination*, then, is the claim that for every fact at time t_0 there is a large number of distinct determinants at all times t_1 where $t_1 > t_0$, but there is no $t_1 < t_0$ for which the fact has a large number of distinct determinants. In a deterministic world the complete state of the world at one time determines the state of the world at any other time, both in the past and in the future. But Lewis claims that there is an asymmetry, in that many disjoint incomplete specifications of the state of the world at t_0 are each individually nomically sufficient for events in the past of t_0, while events in the future of t_0 in general are determined only by the complete state of the world at t_0 (or, in the case of relativistic theories, the complete state in the backward light cone of the event in question). Lewis expresses this in the postscripts to his paper as the claim that many incomplete cross sections of the world postdetermine incomplete cross sections at earlier times, while only complete cross sections predetermine cross sections at later times (Lewis 1986b, 57–58).

Even though the result of Lewis's analysis is that causes generally precede their effects, the temporal direction of causal chains is not a matter of definition. For it may be the case that according to some theory the direction of overdetermination is reversed, or it may turn out that even in our world, which is by and large forward-causal, there are certain pockets of overdetermination of the future by the past, which would imply the presence of backward causation. Thus, Lewis's theory has the prima facie advantage that it does not rule out theories like Dirac's theory of the electron by definition.

The overdetermination thesis, then, is crucial to Lewis's account. Without the asymmetry of overdetermination, Lewis is left without an explanation for the asymmetry of counterfactuals, which in turn is meant to underwrite the asymmetry of causation. Why should we believe the overdetermination thesis? What is striking is that despite its importance to his account of causation, Lewis offers almost no argument for the thesis. He says, as we have seen, that the thesis "is plausible." But there is no general argument for why we should share Lewis's intuitions that the thesis should be plausible, nor is there an argument that could justify his slide from claiming the plausibility of the thesis in the first half of the sentence quoted above to his fully asserting it in the second half.

To the extent that there is any argument, it is one by example. Lewis appeals to his discussion of the counterfactual "If Nixon had pressed the button there would have been a nuclear holocaust" (1986b, 50) and he invokes the asymmetry exhibited by wave phenomena as "a special case of the asymmetry of overdetermination" (ibid.). But the first example is not of much help in trying to assess the thesis, since Lewis relies largely on what he takes to be our shared intuitions about its features and gives us no detailed account of the relevant laws in light of which facts about the future are supposed to be nomically sufficient for present facts. But if we are to accept that "there is no lawful way for the combination [of future traces] to have come about in the absence of the fact," then we ought to look at our best candidates for (deterministic) physical laws and ask if they support the thesis. Lewis's second example is a lot more useful in this respect. In the case of classical wave and radiation phenomena we know what the laws are—the inhomogeneous wave equation that in the case of electromagnetic waves can be derived from the Maxwell equations—and we can expect a unequivocal answer to the question of whether wave phenomena do in fact exhibit an asymmetry of the kind Lewis postulates.

To illustrate the purported asymmetry, Lewis cites Popper's famous example (whose account of the asymmetry we discussed in chapter 5)—the case of circularly diverging waves that result if a stone is dropped into a pond whose water surface was initially smooth (see Popper 1956a). As Lewis describes the case, "there are processes in which a spherical wave expands outward from a point source to infinity. The opposite processes, in which a spherical wave contracts inward from infinity and is absorbed, would obey the laws of nature equally well. But they never occur" (Lewis 1986b, 50).

This, according to Lewis, is an example of the extreme overdetermination of the past by the future, since "countless tiny samples of the wave each determine what happens at the space-time point where the wave is emitted" (ibid.). Lewis appears to believe that philosophical reflection alone can reveal whether wave phenomena do in fact exhibit this asymmetry. Unfortunately, this rather optimistic view about the powers of pure philosophical inquiry is not justified. Lewis's philosophical intuitions about wave phenomena, despite perhaps being initially plausible, are not supported by the physics governing these phenomena.

Lewis says that the asymmetry of radiation is "a special case of the asymmetry of overdetermination" and "that the same is true more generally" (ibid.). But this asymmetry is more than one special case among many; indeed, it may be Lewis's most promising candidate for an asymmetry of overdetermination within the context of classical, deterministic physics. For wave and radiation phenomena at least exhibit *some* form of asymmetry, and Lewis might hope that this asymmetry can serve to underwrite an asymmetry of overdetermination.

3. Liouville's Theorem and the Right Kind of Properties

Frank Arntzenius thinks there is a very general argument that can show that Lewis's thesis of the asymmetry of overdetermination is false (Arntzenius 1990). Arntzenius's argument appeals to Liouville's theorem, according to which the size of any subset of the phase space for an isolated system remains constant under a deterministic (Hamiltonian) time evolution for the system. Possible states of the system correspond to points in the phase space. A deterministic evolution is a 1-to-1 mapping of points in phase space onto points, since each state at one time corresponds to exactly one state at any other time. Liouville's theorem, thus, is a principle of determinism: If each point at t_0 evolves into exactly one point at t_1—that is, if particle trajectories neither split nor merge—then the size of a set of points remains constant under time evolution.

If, with Arntzenius, one identifies properties with subsets of a system's phase space, Liouville's theorem implies that there exists a 1-to-1 mapping of properties onto themselves, such that an isolated system has property p at time t_0 if and only if it has property $F(p)$ at time t_1. The existence of such a mapping, according to Arntzenius, is enough to show that Lewis's thesis of overdetermination is false, since it shows that "there is for any fact at any time a unique determinant at any other time" (Arntzenius 1990, 84). But this argument is a bit too quick.

Arntzenius's notion of property is rather liberal. Thus, it is open to Lewis to reject the 1-to-1 correspondence between properties and subsets of state space postulated by Arntzenius and argue that not every subset corresponds to a genuine property. As is well known from discussions on the statistical foundation of thermodynamics, compact and localized 'blobs' in phase space can evolve into thinly spread-out and highly fibrillated regions (without a change in volume). Lewis need not accept that such regions of phase space represent genuine or 'natural' properties.

In particular, Lewis might argue that the fundamental properties—those which are the subject of the thesis of overdetermination—are spatially localized. Lewis's view appears to be that systems that are localized to some appropriately small region of space at one time t_0 tend to spread out as they evolve in time, and that there can be many localized subsystems of such spread-out systems at a time t_1 that determine facts about the localized past of a system. Or, alternatively, subsystems interact locally at t_0 and move away from one another, where the state of each subsystem at t_1 determines facts about the system as a whole at t_0. Emphasizing this purported feature of the world, one can formulate the overdetermination thesis in a way that is compatible with Arntzenius's challenge: A system's local substates at one time are overdetermined by many maximal local substates of the system at future times (or perhaps by many disjoint sets of such substates), such that the determining substates are states of spatially nonoverlapping and noncontiguous subsystems of the system; but local substates are not similarly determined by substates in the past. Here a maximal local substate is the largest substate of a system localized to some small continuous region of space.

A simple toy example may help to illustrate the view. Imagine a deterministic device that ejects either two red or two green balls in opposite directions. The device can be in one of two states: a red state or a green state. At some time t_1 after the device ejects two balls, the entire system has spread out and consists of three localized and spatially isolated subsystems: the ball on the right, the ball on the left, and the device in the middle. The state of the device at t_0, as it ejects the balls, is overdetermined by each of its subsystems at t_1: For example, if there is a red ball on either side, the device earlier was in the red state. And similarly, we can imagine that there is a record of the ejection event in the device. Each of the substates, such as red-ball-on-right, is a maximal local substate, since the whole system has no larger substates that are localized in that particular region of space. And the entire system is spread out through distinct, noncontiguous subregions of space. Now, even though the later state of the device and the later states of the balls are also overdetermined by the individual states of the balls in the device at the moment of ejection, there is an asymmetry here: The state of a ball at emission is not a maximal local substate, since until after the emission the entire system is localized in the region where the device is located. (Similarly, the top half of the right red ball does not constitute a maximal local substate.) Thus, there are multiple, spatially isolated subsystems at t_1 whose respective states all individually postdetermine the state of the system at t_0, but there are not similarly multiple spatially isolated subsystems at t_0 that predetermine the state of the system at t_1. The states of subsystems at later times serve as distinct 'records' of the state of the system at earlier times.

That it is plausible to construe Lewis's thesis along these lines is, as I suggested above, supported by his discussion of what distinguishes big miracles from small ones. At a point in the discussion where all that is at issue is the size of a miracle, Lewis contrasts a big miracle with one that is "small, localized, simple" (Lewis 1986b, 49). And he says in the Postscript that "what makes the big miracle more of a miracle is... that it is divisible into many and varied parts, any one of which is on a par with a little miracle" (Lewis 1986b, 56). One important way in which miracles are divisible, according to Lewis, is by being spread out through space. Thus, one criterion by which we assess the size of a miracle is how localized it is: "In whatever way events may be spread out or localized, unlawful events can be spread out or localized" (ibid.). Of course, in light of the discussion in chapter 4 we can see that there is an additional hidden and unargued-for assumption in Lewis's claim that spread-out changes in the world require a 'big' or spread-out miracle. For if a miracle's effects were felt instantaneously throughout the world, even a small, localized miracle could have big effects. Creating an additional massive body, for example, affects the motions of all other bodies instantaneously through the gravitational force associated with the body.[3]

In our example, Lewis's intuition would tell us that it would be a smaller miracle to alter the state of the device locally before the balls are released, so that it ejects green balls rather than red ones, than to change the colors of both spatially separated balls after they have been ejected (and to change a record of the earlier ejection in the device). The local change would require a smaller miracle, even though the phase-space regions corresponding to both the earlier local and the later nonlocal property that is changed miraculously have the same size. And this asymmetry of miracles is explained by an asymmetry of overdetermination.

One might worry that even in this amended form overdetermination is still symmetric. For do not local substates at any time after the ejection overdetermine the state of the system at even later times? But, first, the situation is not entirely symmetric, since there is a time—the time just prior to the ejection—at which future states are not overdetermined, but the present state is overdetermined by the future. And, second, since Lewis maintains that the asymmetry of over-determination holds only in very complex worlds like ours, he could maintain that in such worlds there are at each time some (or even many) localized systems that are just about to spread out into spatially isolated subsystems.

But what about the state of the subsystems *before* the time t_0, when the entire system was localized? In general, the system would not have been localized in the remote past. So should we not expect that the states of past localized subsystems predetermine the state of the system at t_0 in the same way the system's future subsystems postdetermine the system's state at t_0? But there are two reasons why one might hold that the situation is not symmetric. First, Lewis might claim that before the system's various components were 'assembled' locally at t_0, the system was not sufficiently isolated. So Liouville's theorem simply does not apply. Relatively isolated subsystems of the world come into existence locally, one might claim, and during their existence as sufficiently isolated systems, their respective subsystems spread out. Of course, in a sufficiently complex world the components of a system will not remain isolated from the rest of the world for very long, but they may still remain sufficiently isolated in certain relevant respects. For example, in our example the two ejected balls might collide with other objects. But as long as there is no mechanism that can change the balls' colors, the subsystems are isolated in the respect that is relevant for them to provide a record of the system prior to the ejection of the balls. Second, even if the system had been isolated all along, a defender of Lewis could once again appeal to the fact that not every subset of phase space corresponds to a genuine property. Thus, it might be the case that, compatible with Liouville's theorem, facts about basic properties of later localized subsystems overdetermine facts about the system at t_0, but facts about basic properties of earlier localized subsystems do not.

Of course none of these considerations show that Lewis's thesis of over-determination is true. In fact, the thesis is false. But the considerations do show, I believe, that merely appealing to Liouville's theorem is insufficient to undermine the thesis. Arntzenius's argument alone cannot disarm what appears to be the core intuition driving Lewis's account, namely, that the states of localized subsystems of a spatially spread-out system individually and multiply determine facts about the system's past.

4. Waves on Popper's Pond

4.1. Overdetermination

I now want to turn to my own criticism of Lewis's thesis of overdetermination. Do radiation or wave phenomena in fact exhibit an asymmetry of overdetermination, as Lewis claims?[4] Is it true that there are many incomplete cross sections that postdetermine incomplete cross sections at prior times (while it is not the case that there are many incomplete cross sections that predetermine incomplete cross sections at later times)? Diverging waves spreading on a pond or the radiation emitted by a source might seem to be the kind of system that begins its life localized and then spreads out. That is, they seem to be the kind of system with distant subsystems whose states might individually determine an earlier, more localized state of the system. As we will see, however, this is not so.

Let us begin by looking at a different and simpler problem than that of a charged particle in an external field—that of free fields without charges. In this case the wave equation for the electromagnetic potential which can be derived from the source-free Maxwell equations specifies a pure Cauchy problem. That is, the field at any space-time point x can be determined from the values of the potential and its derivative on a spacelike hypersurface. Because the Maxwell equations are time-symmetric, the Cauchy data determining x can be specified both on a past and on a future hyperplane. Moreover, since the fields propagate at a finite speed, one does not need to know the fields and their derivatives on the entire hyperplane, but only those in the intersection of the hyperplane with the past or future light cone of x: Incomplete cross sections are sufficient to determine the field at other times. The important point for our discussion is that the problem is temporally completely symmetric: The same kind of 'initial conditions' specified on some spacelike hyperplane are needed both to predict and to retrodict the electromagnetic field at x. In fact, this point generalizes to any theory with time-symmetric laws that allows for the formulation of a pure initial-value problem. Thus, if there is a temporal asymmetry of overdetermination, it must be due to the presence of charged particles or wave sources and to the fact that the resulting problem apparently cannot be solved as a pure Cauchy problem.

If there are charges present, then the specification of the fields on a spacelike surface no longer uniquely determines the fields in the past or the future, since the charges generate additional electromagnetic fields. Recall our discussion in the previous chapters: If we make the standard assumption that the fields associated with a charge propagate with the speed of light away from the charge (in the usual time direction), then the field produced at a point Q propagates along the *future* light cone of Q. In three-dimensional space the field due to a charge at Q has the shape of a diverging sphere whose radius increases at the space of light. That means, conversely, that at a point P the component of the field due to charged particles is determined by the motion of the charges at the points of intersection between their worldlines and the *past* light cone with vertex at P. In this case the field at P is the *retarded field* associated with Q. In the case of a single charged particle in an external field, the total field at P, then, is the sum of the retarded field F_{ret} due to the charge at the retarded point Q and the incoming external fields F_{in}, which may be determined by Cauchy data on a hyperplane that includes Q: $F = F_{ret} + F_{in}$. The problem, thus, no longer is a pure Cauchy problem. In addition to the incoming fields on a Cauchy surface, one needs as input all the worldlines of all charged particles in the space-time region of interest.

As we have seen in the last three chapters, the representation of the total field as sum of a free incoming field and a retarded field is not unique. One can equally represent the field in terms of a final-value problem, as the sum of an *advanced* field (that represents a field concentrically converging into the charge) and a free (outgoing field), $F = F_{adv} + F_{out}$. But the situation is not symmetric, since generally it is easy (in situations in which we are interested) to pick a hypersurface on which incoming fields are zero, but outgoing fields will in general be nonzero. According to the view that I defended in the last chapter, this asymmetry is due to the fact that each charge physically contributes a component to the field that can be represented as a *purely* retarded field, but not as a *purely* advanced field. (Recall that on that view, an advanced representation of the total field, while mathematically possible, is physically somewhat misleading: The retarded field physically associated with a charge is represented as a combination of an advanced field and a *source-free field*.) The question now is whether this asymmetry, whatever its correct explanation might be, gives rise to an asymmetry of overdetermination.

Price has argued against Lewis that radiation phenomena do not exhibit an asymmetry of overdetermination on the microscopic level. This is so, according to Price, because the asymmetry of radiation "is an essentially macroscopic asymmetry" (Price 1992, 505). Price's argument crucially depends on his own account of the asymmetry of radiation, according to which radiation phenomena are symmetric on the micro level and the asymmetry is due to the de facto condition that there are macroscopic coherent emitters but no macroscopic coherent absorbers. I have criticized this account in chapter 6. For my present purposes this disagreement with Price is unimportant. Neither of us believes that radiation exhibits an asymmetry of *overdetermination* on the micro level. But as far as the macro level is concerned, Price concedes too much to Lewis, for there is no asymmetry of overdetermination on the macro level either.

In order to assess Lewis's claim that the radiation associated with a charged particle exhibits massive overdetermination of the past by the future, but not vice versa, we need to ask, first, whether there are multiple sets of distinct, lawfully sufficient conditions specified on a hyperplane in the future that determine the motion of the charge at Q; and, second, whether there are no (or at least significantly fewer) multiple sets of sufficient conditions specified on a hyperplane in the past of Q that determine the motion of Q. Now, one can see immediately that Lewis's claim that "countless tiny samples of the wave each determine what happens at the space-time point where the wave is emitted" is false. For the field at any point P on the future light cone of Q is the *total* field, which in general includes an arbitrary, nonzero source-free field and, thus, the field at P does not even allow us to uniquely determine the contribution to the field at P due to the charge at Q. All that is determined by the tiny sample of the wave at P is the total field strength at this point, not to what extent this field is due to charges in a certain region or to source-free incoming fields.[5]

What if we assumed in addition that there is no free field and that the entire field is due to the motion of the charge? There are two possible ways in which one might introduce this assumption. First, it might be thought of as specifying the initial conditions for a (partial) Cauchy problem. The condition then states that the fields are zero everywhere on some spacelike hyperplane that constitutes the *past* boundary to the space-time region we are focusing on. In the case of Popper's pond, this is the assumption that the surface of the water initially is still. But notice that this assumption cannot provide support for Lewis's thesis, since what Lewis claims is that there are sets of conditions specified *entirely on future cross sections* which determine the past. The assumption of zero incoming fields is instead an assumption about the *past*. Thus, on this construal Lewis could at most show that, given the laws *and certain facts about the past of the world*, facts about the future overdetermine the present.

Alternatively, we could maintain that all fields are ultimately due to charged particles somewhere. If we then focus on a world with just a single charge and assume that the retardation condition holds at that world, then the total field in that world is the retarded field associated with that charge. Thus, in such a world, the "tiny sample" of the field at P is entirely due to the motion of the charge when it intersected the past light cone of P. The assumption that the incoming field is zero on that picture is a general assumption about the type of world in question, and not an assumption about the state of the world at some time in the past.

The problem, however, is that even then the field strength at P is not a "postdeterminant" for any subregion of any particular spacelike hyperplane in its past. The retarded field due to a charge is a complicated expression involving the distance between the field point and the charge, and the velocity of the charge and its acceleration. It can be written as follows:

$$E(x, t) = e\left[\frac{\mathbf{n} - \boldsymbol{\beta}}{\gamma^2(1-\boldsymbol{\beta}\cdot\mathbf{n})^3 R^2}\right]_{ret} + \frac{e}{c}\left[\frac{\mathbf{n}\times\{(\mathbf{n}-\boldsymbol{\beta})\times\dot{\boldsymbol{\beta}}\}}{(1-\boldsymbol{\beta}\cdot\mathbf{n})^3 R}\right]_{ret}, \quad (8.1)$$

$$\mathbf{B} = [\mathbf{n}\times\mathbf{E}]_{ret}$$

where **E** and **B** are the electric field and the magnetic field, respectively; **n** is a unit vector pointing from the point P to the retarded source point Q; R is the distance between P and Q; $\boldsymbol{\beta} = \mathbf{v}/c$; and $\gamma = \sqrt{1-\beta^2}^{-1}$ (see Jackson 1975, eqs. 14.13, 14.14). All quantities are evaluated at the retarded position of the charge. The details of this equation are not important for our purposes here. What is important is that, as inspection can easily show, the value of the field at P alone uniquely determines neither where its source was located nor how fast the source was moving or accelerating. Since the unknown variables—the size of the charge, its relative position (which determines R and $\mathbf{n}$), its three-vector velocity, and its three-vector acceleration—can all be varied independently, knowledge of the fields at P leaves the state of the charge at the emission event radically underdetermined. Moreover, it is of course also impossible to calculate the values of the variables from four arbitrary field measurements (which would result in twelve equations, one for each of the directional components of the fields), since the field at different points will in general be associated with different retarded source points.

As the expression above suggests, the field associated with the charge can be divided into two components, a generalized Coulomb field which is independent of the acceleration of the charge and falls off as $1/R^2$, and an *acceleration field* which varies as $1/R$. Only the latter field carries energy away from the charge. One can show that a charge radiates energy if and only if it is accelerated. But one cannot even tell from the field strength at a single point P alone whether the field in question includes a radiation field and, hence, whether the charge emitting the field is accelerated (see Rohrlich 1965, sec. 5.2). There is no strictly local criterion of radiation that gives us sufficient conditions for the presence of radiation in terms of the field strengths at a single space-time point. As Rohrlich argues, in order to determine whether the field in question is due to an accelerated charge, one could do one of two things. One could measure how the field drops off along the future light cone of the emission event far enough from the source. Sufficiently far away the Coulomb field becomes negligible compared with the radiation field. So if the field is due to a radiating charge, it should vary as $1/R$ in that region. But clearly this criterion is of no help for Lewis, since it does not involve only the state of the field at a single time. Or, if one somehow knew the location and velocity of the charge during the emission, one could measure the fields on the surface of the *entire* diverging light sphere. From this one can calculate the *radiation rate*. The radiation rate is nonzero exactly if the charge was accelerating when it passed through the center of the sphere. Thus, not even the fact that the charge is accelerating (let alone its exact state of motion) is overdetermined by local field values.

The upshot of this discussion is that even with the assumption of no free fields, "tiny samples" of the field do not individually determine the state of the source of the field. And even knowledge of the field on the entire sphere with radius ct centered at the point of emission does not by itself determine the state of the charge. But perhaps less than full knowledge of the state of the charge at emission is sufficient for Lewis's claim that past events leave many traces in the future, but that future events are not multiply foreshadowed in the past. Perhaps it would be enough for Lewis that there are some properties of the system in the past that are overdetermined by the future. But what could these be? We have just seen that not even the fact that there was an accelerating charge at a certain location is overdetermined by future field excitations. A fact about the past that *is* determined by a nonzero field at P (if we assume no free fields) is that the trajectory of a charge *somewhere* intersected the past light cone with vertex at P. Nonzero field values at different points, then, are all individually sufficient for the presence of the charge in the junction of their past light cones. That is all. The overdetermination here is far less than that asserted by Lewis, for there is no event whose time of occurrence is determined by the nonzero fields at P.

Is there an *asymmetry* of overdetermination in this attenuated sense? No, there is not. Since the speed of charged massive particles is less than that of light, a particles that intersects the past light cone of P will also intersect the future light cone of P. Thus nonzero fields at different points are also individually sufficient for the presence of the charge somewhere in the junction of their future light cones. This inference relies on nothing more than the principle of charge conservation, which is a consequence of the Maxwell equations.

I have focused on Lewis's claim that each tiny wave sample individually determines the state of the source. But the overdetermination thesis also allows for disjoint *sets* of conditions as determinants. It is not difficult to see, however, that this is of no help to Lewis here. There is no general overdetermination of the past by the future even in simple worlds with only a single charge, and there is no overdetermination at all in complex worlds like ours.

But, one might ask, does not our world, with its many charges and perhaps source-free fields, exhibit situations like Popper's pond, where a single charge alone seems to be responsible for the excitations of the electromagnetic field in a finite spatiotemporal region? And, thus, is not Lewis's overdetermination claim true in situation like these? But (to repeat) the problem is that in a world with multiple charges and source-free fields, individual wavelets determine certain facts about the wave's source only in a certain region of space-time under the assumption that the incoming field at the past boundary of the region is zero. Thus, in such a world no facts about the future *alone*, but only facts about the future *in conjunction with facts about the past* of the emission event, determine some facts about that event. To put it another way, in a world with free fields it would not have to be necessary that the field be nonzero at P for the trajectory of some charge to intersect the past light cone of P. Of course we can specify special circumstances such that *given these circumstances*, it is necessary for a charge to have intersected the past light cone of P. But these circumstances have to be specified in terms of the past of the emission event, and not its future.

I have discussed this case in some detail to show that philosophical armchair reflections could on occasion benefit from a glance into a physics textbook. Lewis's thesis of overdetermination, I have shown, is false for radiation phenomena. The conclusion we should draw from this, as I have suggested above, is that Lewis's thesis fails in general. For the asymmetry of wave phenomena provides us with what is quite plausibly the best candidate for an asymmetry in classical deterministic theories. Wave phenomena satisfy a genuinely asymmetric constraint — the retardation condition. But other classical deterministic theories do not contain any such asymmetric constraints. As we have already discussed, the asymmetry thesis fails even more obviously for charge-free electromagnetism. But the same reasoning that applies there applies to classical particle mechanics with its time-symmetric laws. The set of coupled differential equations for a mechanical n-particle system defines a pure Cauchy problem: The temporal evolution of the system is determined by the initial positions and velocities of the system, where it does not matter whether the conditions are 'initial' or 'final.'

4.2. Miracles

What are the consequences of the failure of the asymmetry overdetermination for Lewis's theories of causation and counterfactuals? On Lewis's account, the asymmetry of overdetermination is meant to explain an asymmetry of miracles: "[W]hat makes convergence take so much more of a miracle than divergence, in the case of a world such as [ours], is an asymmetry of overdetermination at such a world" (Lewis 1986b, 50). But Lewis might be wrong about this, and perhaps there still could be an asymmetry of miracles even in the absence of an asymmetry of overdetermination. A radiating charge does, in some intuitive sense, leave a variety of increasingly spread-out traces—the charge does have an effect on the state of the field at a multitude of locations—even if these traces are not individually sufficient for the charge to have radiated. All these traces would somehow have to be manufactured in a world that converges to the actual world, but one in which the charge did not radiate. And one might think that manufacturing these traces would require a larger, more spread-out miracle than one that would be sufficient to lead to a diverging world.

There is in fact an extremely simple and general argument showing that there cannot be an asymmetry of miracles in a world with time-symmetric laws. Recall how we construct a world diverging from ours: We introduce a small change to the present state of the world and then let the resulting counterfactual instantaneous state evolve forward in time in accord with the laws of the actual world. But of course we can construct worlds converging to ours through a small miracle by a strictly analogous procedure: We introduce the same small change to the present state of the world and then let the resulting counterfactual present state evolve *backward* in time in accord with the laws of the actual world. By construction, the miracles are the same, and hence of the same magnitude.

While this argument proves that there is no asymmetry of miracles, I still think it is useful to look at how the relevant construction proceeds in detail in the case of classical electrodynamics. For the intuition that the asymmetry of wave phenomena must lead to an asymmetry of miracles is powerful. Since waves *spread*, how can it be that a small miracle can remove all the traces a wave source has left? Showing exactly where these intuitions go wrong can help to guard against similar mistakes in a way in which the general argument cannot.

What, then, does classical electrodynamics have to say about the question of what sorts of miracles are required for divergence or convergence of worlds, respectively? In our discussion of Arntzenius's challenge above, we noted that one important criterion for assessing the size of a miracle is how spread-out or localized it is. A localized miracle is one that changes the state of the world only locally. But then we run into an immediate problem with trying to implement Lewis's framework in the context of classical electrodynamics and compare the sizes electromagnetic miracles. For some of the obvious candidates for small Lewisian miracles could not possibly be merely local miracles in classical electrodynamics.

The Maxwell equations impose not only dynamical constraints that govern how the state of a system changes, but also synchronous constraints on what configuration of charges and fields is possible at a moment. That there are such additional constraints can easily be seen by looking at the order of the equations:

The electromagnetic field has six degrees of freedom at each point—the components of the electric and magnetic field vectors—while the Maxwell equations provide us with eight independent conditions as constraints: the two equations for the divergence of the electric and magnetic fields, and the two vector equations, which are equivalent to six equations relating the individual components of the fields to components of the curl of the field. In particular, the synchronous constraints imply that each charge is surrounded at all times by its Coulomb field. Thus, we could not locally create or destroy a charge through a 'small' miracle without affecting the field everywhere. What is more, we could not even miraculously change the location of a charge without affecting the field everywhere.

One might think that once we introduce the notion of a miracle violating the dynamical constraints imposed by Maxwell equations, it is only a small step to allowing miracles that violate synchronous constraints as well. Thus, for example, we should be allowed to consider a miracle creating a new charge without instantaneously affecting the electromagnetic field everywhere. But it is not clear how we could then use the Maxwell equations with a contralegal state as input to generate well-defined predictions or retrodictions. The theory is silent on what the past or the future evolution of a world with a contralegal present would be, and we therefore could not implement Lewis's prescription for determining the temporal evolution of such a world, according to which the world evolves according to the laws of the actual world both after and before the miracle.

If even intuitively 'small' miracles are necessarily spread through all of space, how we should compare the sizes of miracles is even less clear than it might have appeared originally. Does it, for example, take a smaller miracle to create a single very large charge, which has relatively noticeable immediate effects on the total field even at large distances, than to create two very small, yet far-apart charges, which affect the total instantaneous field only minimally? Since this is a problem not only in classical electrodynamics, but in the Newtonian gravitational theory with its instantaneous action-at-a-distance force as well, this raises serious doubts about whether Lewis's notion of a ranking of the size of contralegal miracles can be spelled out in any meaningful way.

There is, however, one type of miracle that appears to be fairly unproblematic from the standpoint of electrodynamics—miracles that change the instantaneous acceleration of a charge. Miraculously changing the acceleration of a charge now (perhaps by hitting the charge with a supernatural microscopic hammer) has no effect on the field now but, through a change in the trajectory of the charge, will have an effect on the field at later or earlier times (depending on whether we set the situation up as an initial- or a final-value problem). So I would like to consider an arbitrarily complex world w_0 (a world like ours) consisting of charged particles and fields, and I want to ask whether there is a difference in size between convergence and divergence miracles, restricting myself to miracles that change the trajectory of charges by changing the accelerations of charged particles. Does convergence require a larger miracle than divergence?

At this point (after having labored so hard to understand the structure of classical electrodynamics) it should be clear that the answer is 'no.' We begin by specifying the state of the world (consisting of charges and fields at one time).

180 *Inconsistency, Asymmetry, and Non-Locality*

Then we can ask how the future would be different if the acceleration of a single charge would have been different. The resulting problem is an initial-value problem. The total field at some space-time point in the future of the initial-value surface at t_0 is given by $F = F_{ret} + F_{in}$, the sum of a free field and the retarded field due to all the charges. (Recall that F_{in} is a solution to the homogeneous wave equation that agrees with F on the initial-value surface.) If one of the charges has a different acceleration now, its trajectory in the future will be different, and hence the retarded field associated with the charge will in general be different. A small miracle leads to a world w_d that diverges from w_0. Also, even though the miracle is small, it has widespread effects even in very simple worlds containing only a single charge. For the difference in field excitations will affect the field farther and farther away from the charge at later and later times.

But of course the situation is entirely symmetric, in that two worlds with different pasts could be made to converge through the same kind of small miracle. Now we are considering a world in which a charged particle has a trajectory different from that in the actual world in the past—a trajectory that, according to the laws, would cross the actual trajectory at the time when we want to introduce the miracle. Due to a miraculous change in acceleration, the charge's trajectory is then made to converge with the actual trajectory. To find out what the different pasts of these worlds are, we have to solve a final-value problem. By assumption, both worlds share the same future and thus agree not only on the trajectories of the charges but also on the fields from the moment of the miracle onward. Thus the field at a point in the past of the time of convergence is given by $F = F_{adv} + F_{out}$, where F_{out} by assumption is the same field as in the actual world. The difference between the pasts of the two worlds—the world w_0 and a world w_c that converges to it—is due to the difference in the trajectory of the charge whose acceleration was changed and the associated advanced field.

Lewis asks, "How could one small localized, simple miracle possibly do all that needs doing?" (Lewis 1986b, 49). How can it be that the total fields agree at t_0, even though the charges which are (at least partly) responsible for the field had different past trajectories in the two worlds? Let us look at the converging worlds w_0 and w_c and ask how the field at t_0 comes about in the two cases. In w_0 the field at t_0 at some point P is the sum of all the retarded fields associated with all the charges and any free incoming fields. (We can calculate the field at t_0 in terms of an initial-value problem set up at an earlier time t_{-1}.) How does the same field come about in the miracle world w_c? Since the trajectory of one of the charges in this world is different, the difference in the resulting retarded field has to be made up for by a difference in the free field. And in fact we can easily write down what the difference in the free field has to be. We have seen in chapter 6 that the field $F_{ret}^i - F_{adv}^i$ for a charge i is a solution to the homogeneous Maxwell equations. This field looks like a wave collapsing inward onto the location of i and then diverging outward again. Thus, if the charge i in w_0 has a different trajectory (which I want to label i^*), then the free field in w_c, $F_{in,c}$, has to differ from any free incoming field in the actual world, $F_{in,0}$, by

$$F_{in,0} - F_{in,c} = (F_{ret}^i - F_{adv}^i) - (F_{ret}^{i^*} - F_{adv}^{i^*}). \tag{8.2}$$

The net effect of this field at t_0 (and later) is to cancel the retarded field associated with the charge i in w_c and add a contribution that looks like the retarded field associated with i in w_0. Thus, due to a 'cleverly designed' free field, in the miracle world w_c at t_0 it looks as if its past had been that of w_0.

My construction of a world that converges through a small miracle may remind some of Lewis's discussion of what he calls a "Bennett world" (Lewis 1986b, 56–58). A Bennett world is obtained as follows. We begin with a world w_0 like ours and introduce a small miracle to arrive at a diverging world w_1. Then we extrapolate the latter part of w_1 backward in accordance with the laws. The result is a Bennett world, which is free of miracles and agrees with w_1 after the miracle that made w_1 diverge from the 'base' world w_0. Importantly, it takes only a small miracle for w_1 to converge with the Bennett world—the same small miracle that was required for w_1 to diverge from w_0.

Lewis's response to the problem presented by the existence of Bennett worlds is to argue that a Bennett world is not like ours. Unlike our world, a Bennett world is one to which convergence is easy. Lewis claims that this difference is related to the existence of what he takes to be de facto asymmetries in our world, such as the radiation asymmetry—asymmetries which he says must be absent in a Bennett world. Ultimately, such differences, according to Lewis, can be accounted for by appealing to the asymmetry of overdetermination: In worlds like ours, there are "plenty of very *incomplete* cross sections that postdetermine incomplete cross sections at earlier times. It is these incomplete postdeterminants that are missing from the Bennett world" (Lewis 1986b, 57–58). In a Bennett world only a complete cross section "taken in full detail, is a truthful record of its past history" (ibid.).

Now since Lewis's overdetermination thesis is false, this cannot be what distinguishes Bennett worlds from worlds like ours. More important, however, my example of a converging world shows that the construction of a Bennett world is unduly complex and that it serves only to obscure the issue. For in the case we discussed here, the world to which the miracle world converged is simply a world like ours. Thus any attempt to appeal to putative differences between a Bennett world and ours is of no help here. It takes no special world for convergence to be easy. Take the state of the world at a certain time t_0. We get a world diverging to the actual world if we introduce a miracle at $t_0 + \varepsilon$ and then run the laws forward. Alternatively, we get a world converging to the actual world if we introduce *the very same kind of miracle* at $t_0 - \varepsilon$ and run the laws backward. There is no difference in the size of miracles required.[6]

4.3. Amending the Similarity Metric

I have shown that Lewis's account of an asymmetry between worlds converging to and diverging from the actual world is untenable. Contrary to what Lewis claims, there is no difference in the size of the miracles required to achieve perfect convergence and perfect divergence. Moreover, neither do converging and diverging worlds differ with respect to the extent to which they disagree with the actual world about particular matters of fact that hold in these worlds. That is, neither of Lewis's original standards of similarity allows us to discriminate between converging and diverging worlds. For both kinds of worlds need contain only a small and isolated violation of the laws that hold in our world, and both kinds of worlds disagree widely with the actual world on particular matters of fact. Just as the fields differ in the future in a diverging world, so past fields are different in a converging world. And, as we will see in more detail below, in both cases the electromagnetic fields differ from those in the actual world by precisely the same amount.

While the world to which convergence is easy is just a world like ours, there is, however, something peculiar about worlds that, like w_c, converge to ours as the result of a small miracle. In such a world, as we have just seen, there need to be carefully set-up, source-free incoming fields that, in a sense, serve either to mimic the presence of charges or to conceal their presence. Of course, charges are not really concealed, since in that world, as in ours, each charge contributes to the total field in the same way. But what the need for such compensatory fields implies is that in the converging world w_c the Sommerfeld radiation condition of zero free incoming fields, $F_{in} = 0$, cannot in general be satisfied. Moreover, the incoming fields will contain coherent disturbances that cannot be mathematically associated with any source. There is no single charged particle that can be associated with (either the retarded or the advanced component of) the contribution $F^i_{ret} - F^i_{adv}$ to the free incoming field—the component of the total field that in w_c mimics the trajectory of a in w_0, since the relevant charge has trajectory i^* (and not trajectory i) in w_c.

This suggests that there might be a way in which a broadly Lewisian account of the asymmetry of causation and of counterfactuals might be defended. Lewis could, it seems, amend his similarity metric to include a provision that worlds exhibiting coherent field disturbances which cannot be associated with any source are much less similar to the actual world than worlds not containing such 'inexplicable' fields.

Moreover, the peculiar incoming fields in w_c also have consequences for the thermodynamic behavior of this world. In a world without absorbers, the incoming fields would simply have to be 'set up' at past infinity to lead to perfect convergence with the actual world. The actual world, however, contains absorbing media. In the actual world, absorbing media behave thermodynamically: Incoming radiation is partly absorbed and partly scattered incoherently, such that the entropy of the system increases. In a world converging to ours the very same media will, of course, be present, but they there need to act as anti-absorbers. For example, the particles in the walls of the laboratory in which the miracle occurs need to begin oscillating coherently prior to the miracle in order to produce the field given by (8.2). Recall the discussion in chapter 6: If we evolve the state of a system backward in time in accord with the micro laws, a nonzero field in the absorber's future will correspond to correlated motions at the absorber's future boundary, which get damped in the past direction. In the normal time sense, this looks like spontaneously arising correlations between the 'absorber' particles. Of course, after the miracle, the converging world looks like ours. The charge i now has the same trajectory as in our world. And the total field 'looks as if' the charge always had the same trajectory as i and the walls had never radiated.

This suggests a second way in which Lewis's similarity metric might be amended: In addition to (or perhaps instead of) the requirement that there be no coherent fields not associated with sources, Lewis could require that worlds close to the actual world exhibit the same thermodynamic behavior as the actual world. The anti-thermodynamic behavior exhibited by an absorber in the converging world does not violate the laws of the actual world, but, one could maintain, the behavior is extremely improbable. In fact, it is so improbable that as a matter of fact we never observe such anti-thermodynamic behavior in the actual world. Now, one needs to be a little careful here. For while the anti-thermodynamic behavior of absorbers in the miracle's past is improbable, given the absorbers' low-entropy initial states, the behavior is *not* at all improbable given the state of w_c when the miracle occurs. In fact, given the miracle world's present, the anti-thermodynamic behavior is to be expected! (And neither is the behavior improbable if we conditionalize *both* on the present and on a low-entropy past state of the absorbers.)

Nevertheless, one might argue that we ought to include agreement in phenomenological regularities (such as the second law of thermodynamics) in a similarity metric between worlds. According to the thus amended metric, perfectly converging worlds are much less similar to the actual world than perfectly diverging worlds, since they contain 'inexplicable' coherent radiation and there are phenomenological regularities which are de facto exceptionless both in the actual world w_0 and in diverging worlds w_d (even though violations of these regularities are not nomically excluded), but which are not without exception in converging worlds w_c. Thus, while Lewis is mistaken in trying to account for an asymmetry of counterfactuals in terms of an asymmetry of miracles or an asymmetry of overdetermination, there may yet be a similarity metric that makes backtracking counterfactuals come out false in general.

The question, then, is whether amending the similarity metric in this fashion can help to underwrite Lewis's analyses of counterfactuals or of causation. I believe that at least in the case of the analysis of causation, the answer has to be 'no,' for reasons I want first to summarize and then to explain in a little more detail. The amended similarity metric is intimately linked to the wave asymmetry between prevailing initial and final conditions: Converging but not diverging worlds contain coherent radiation fields not associable with any single source exactly if the actual or 'target' world exhibits the radiation asymmetry. In fact, the two asymmetries can be shown to be mathematically equivalent. But then the Lewisian account faces the following dilemma. Either the amended similarity metric is introduced as a basic or fundamental feature of our world that is not capable of being explained further—but then the radiation asymmetry also has to be posited as basic and inexplicable—or one accepts that the latter asymmetry can be explained. But, as I argued in the last chapter, the best explanation of the radiation asymmetry appeals to an asymmetric causal or counterfactual constraint. Thus, the amended Lewisian account becomes circular. The radiation asymmetry is explained by appealing to an asymmetric constraint—the retardation condition—that on the Lewisian account in turn is accounted for by appealing to a difference between the fields in converging and diverging worlds that is formally equivalent to the radiation asymmetry.

Now to the details. Why does the converging world, but neither the diverging world nor the actual world, contain coherent radiation not associated with any

184   Inconsistency, Asymmetry, and Non-Locality

charge behavior different from w_0? What accounts for this difference is the fact that incoming fields in the actual world contain no coherent disturbances centered on sources in their future, while outgoing fields do contain disturbances centered on sources in their past. Contrast w_c with the diverging world w_d. Just as w_c contains an *incoming* field different from that in our world, so w_d contains an *outgoing* field different from that in our world. If the incoming field is zero, then the total field on a hypersurface in the future of the miracle will be $F^{i^*}_{ret}$, instead of F^i_{ret}, for a single charge i. Thus, if we expressed the field in the past of that hypersurface in terms of a final-value problem $F = F_{adv} + F_{out}$, the free outgoing field in the miracle world, $F_{out,d}$, would differ from that in our world by

$$F_{out,0} - F_{out,d} = (F^i_{ret} - F^i_{adv}) - (F^{i^*}_{ret} - F^{i^*}_{adv}), \tag{8.3}$$

which is just the same as the difference in incoming fields (8.2) in the case of w_c. Since, however, the incoming field is zero and the outgoing field is equal to the sum of the retarded fields of all charges, the symmetry between diverging and converging worlds is broken. For then the free incoming in w_c is given by

$$F_{in,c} = (F^{i^*}_{ret} - F^{i^*}_{adv}) - (F^i_{ret} - F^i_{adv}), \tag{8.4}$$

where the second term on the right represents a nonzero coherent field not associated with any charges in w_c, while the free outgoing field in w_d is

$$F_{out,d} = (F^{i^*}_{ret} - F^{i^*}_{adv}). \tag{8.5}$$

Thus, if the incoming field is zero in the actual world, then converging worlds, but not diverging worlds, contain coherent fields not associated with any single source.

Conversely, if there are no coherent fields not associated with any single source in a diverging world, then the first term on the right-hand side in (8.3) has to be canceled by $F_{out,0}$. And this is the case if the outgoing field is purely retarded—that is, if the free incoming field is equal to zero.

Thus, the asymmetry between initial and final fields characterizing radiation phenomena can alternatively be expressed as follows:

> Worlds that agree with the actual world on the present fields, even though charges in those worlds had different past trajectories, need to contain coherent radiation not associable with any single source, while worlds that agree with the actual world on the present fields, but where only the future trajectories of the charges differ, do not contain coherent radiation not associable with any single source.

If we agree with the claim that the asymmetry between initial and final conditions captured by RADASYM in chapter 5 is in need of an explanation, then we cannot simply postulate the amended similarity metric appealing to the presence of 'inexplicable' coherent fields as fundamental, since the two asymmetries are equivalent. Yet if one grants that the radiation asymmetry can be explained, and if one were to accept, further, that the explanation of the radiation asymmetry in terms of a fundamental causal constraint is correct, then an appeal to the amended similarity metric as grounding a Lewisian analysis of the counterfactual and causal asymmetries is in danger of being circular.

At this point we need to distinguish between Lewis's analysis of the asymmetry of causation and that of counterfactuals. Lewis's analysis of causation clearly fails. Since we are appealing to an asymmetrical causal constraint to explain the asymmetry between incoming and outgoing fields, we cannot then explain the asymmetry of the causal constraint by appealing to an asymmetry that is formally equivalent to the asymmetry of the fields.

We can, however, accept an amended Lewisian analysis of the asymmetry of counterfactuals. As long as we reject Lewis's counterfactual analysis of causation, the amended similarity metric can be justified by appealing to the retardation condition as causal constraint without threat of circularity. But this would have the consequence of turning Lewis's account on its head (or, rather, right side up): The truth conditions for counterfactuals would be given in part by appealing to causal features of the world. Worlds diverging from ours are less similar to the actual world than worlds converging to it because of asymmetric causal features of the actual world.

5. Do Independence and Determinism Imply Overdetermination?

Daniel Hausman has argued that the asymmetry of overdetermination is a consequence of two perhaps relatively innocuous assumptions to which Lewis appears to be committed (Hausman 1998). The first assumption is the *independence condition* I: "If a causes b or a and b are causally connected only as effects of a common cause, then b has a cause that is distinct from a and not causally connected to a" (Hausman 1998, 64). Two distinct events, for Hausman, are causally connected if and only if either one is the cause of the other or if they are effects of a common cause.

The second condition is that causes are INUS conditions for their effects: A cause is an *i*nsufficient but *n*ecessary part of a set of *u*nnecessary but *s*ufficient conditions for the effect. Hausman calls this condition *deterministic causation*, DC. Lewis, of course, rejects a regularity account of causation, but his analysis implies that an INUS account is correct in circumstances where there is no 'ordinary' causal overdetermination—that is, in circumstances where there is no alternative possible cause waiting in the wings to bring about the effect if its actual cause fails to operate. And it is in these circumstances that, according to Hausman, one can derive the asymmetry of overdetermination, which in his version reads as follows: "**AOD**: If causation is deterministic, then (1) events will be determined by a great many of their (natural) effects, and (2) events will not be determined by any of their (natural) causes" (Hausman 1998, 136). Since Hausman is not interested in giving an analysis of causation that involves the notion of overdetermination, his version of the thesis can, unlike Lewis's version, invoke the notion of cause.

Hausman's result appears striking in light of the conclusion of the discussion so far. If, as I argued, the thesis of overdetermination is false, then Hausman's argument implies that either causes are not, even in the most 'ordinary' of circumstances, INUS conditions for their effects, or events cannot have distinct causes that independently 'conspire' to bring about the effect. But Hausman's proof of overdetermination is flawed.

Before showing where the proof goes wrong, I want to point out that Hausman's result would be striking for a second reason: The proof purports to derive an asymmetry thesis from premises that are not asymmetric. Both **I** and **DC** are *non*symmetric in the sense that they specify conditions which causes have to satisfy relative to their effects, but do not also demand that the 'causal inverses' of the conditions should hold, where the causal inverse of a condition is obtained by interchanging causes and effects everywhere. But neither condition implies the falsity of its causal inverses: The causal inverse of **I** says that causes have multiple effects (which are not themselves related as cause and effect), and the causal inverse of **DC** says that effects are INUS conditions of causes. Neither of these two conditions is excluded by **I** or **DC**. Yet the thesis of overdetermination excludes its causal inverse—the thesis is asymmetric and not merely nonsymmetric—since it states that effects overdetermine their causes, *but not vice versa*. Thus Hausman's derivation of **AOD** conflicts with the following rather plausible general principle: "No asymmetry in, no asymmetry out." Either this principle is false or Hausman smuggles an asymmetry into his proof somewhere.

To see where Hausman's proof of **AOD** goes wrong, I want to quote it in its entirety:

> Proof: Given **DC** and the absence of ordinary preemption or overdetermination, the conjunction of (the properties of) the causal tropes is necessary in the circumstances for the effect trope, and so each cause is determined by each of its effects. Suppose that a causes b. By **I** there will [be] another cause f of b that is causally independent of a. Since a will not be necessary in the circumstances for f nor vice versa, a by itself—that is without f—will not be sufficient in the circumstances for b. Effects will not be determined by any of their individual causes. (Hausman 1998, 136)

Now, in section 4 we have seen that a local field excitation only (partially) determines the state of its source, if we assume as initial condition that there are no free fields. Hausman's result similarly relies on an appropriate choice of 'initial conditions,' which introduces the asymmetry. He begins by stating, entirely correctly, that if causes are INUS conditions, then *in certain circumstances*, causes are necessary for their effects. Since INUS conditions are *not* necessary for their effects *tout court*, the conditions we have to conditionalize are those in which alternative sets of sufficient conditions, which do not include the cause in question, are absent. In such circumstances the cause will still not be sufficient, so there is an asymmetry of overdetermination. But this asymmetry is entirely due to the way we pick the circumstances that are held fixed, and can be reversed by a different choice of circumstances. For example, in the case of radiation, if we are given the 'outgoing' field as initial condition—that is, the field on a spacelike boundary in the future of the space-time region we are interested in—then the state of a charge at Q will be partially determined (in the ways discussed in section 4) by the field values at field points on the past light cone with vertex at Q. Thus, depending on whether we are given initial or final conditions, the state of the source is partially either predetermined or postdetermined multiply by the field on the charge's light cone.

Hausman's mistake is that the qualification to special circumstances drops out in his conclusion. What can in fact be derived from his two premises is not **AOD**, but rather the following: "If causation is deterministic, *and given the absence of alternative sufficient conditions concerning alternative sets of causes for their effects*, then (1) events will be determined by a great many of their effects, and (2) events will not be determined by any of their causes." But this principle is no longer asymmetric and, like the assumptions from which it is derived, is perfectly compatible with its causal inverse.

We can use an old staple of the literature on causation and explanation to illustrate how the asymmetry depends on the choice of circumstances: the case of a flagpole of a certain height being the cause of a shadow of a certain length. We can specify circumstances in which the causal inverse of the overdetermination thesis is true. For given the right circumstances (such as the sun's being at a certain angle and the absence of clouds or other interfering objects), the presence of a flagpole of a certain height is sufficient for the presence of a shadow of a certain length. But since every event has multiple effects (the inverse of **I**), and the event may be determined only by the conjunction of its effects, the shadow alone is not sufficient for the presence of the flagpole. For example, a shadow of the same length may be caused by different objects, say different flagpoles of different heights at appropriately different locations. Then only the conjunction of effects—including, for example, the effect the flagpole has on the ground underneath it—will be sufficient for the presence of the flagpole of a certain height.

And of course this scenario is compatible with both of Hausman's assumptions, **I** and **DC**: The presence of the flagpole is an INUS condition for the presence of the shadow and the presence of the shadow has other causes besides the flagpole that are causally independent of the presence of the flagpole (such as the sun's being at a certain elevation). Thus, the scenario is also compatible with there being circumstances (of course different from the ones above) such that in those circumstances the presence of the shadow is sufficient for the presence of the flagpole.

6. Convergence Made Easy: The Challenge from Statistical Physics

While I have examined the thesis of overdetermination and that of the asymmetry of miracles in the context of classical electrodynamics, Adam Elga has argued that "in many cases that involve thermodynamically irreversible processes, Lewis's analysis fails" (Elga 2001, S314). Elga argues that slight miraculous changes in a system's final state can lead to radically different histories for the system. A system with a normal thermodynamic past during which the system's entropy never decreased can, through a small localized change in the system's microscopic final state, be transformed into a system which, if we run the laws backward, can be shown to have had an anti-thermodynamic past during which its entropy decreased dramatically.

Elga's picturesque illustration of his argument is the cooking of an egg. In our world an egg is cracked and then cooked in a pan. Elga argues that only slight changes in the micro state of the cooked egg in the pan result in a past for the egg in which the egg slowly formed out of rot and never was cracked. The multiple traces that the cracking of the egg might have left in the actual world, Elga claims (without much discussion of the claim), are all similarly formed in the miracle world through anti-thermodynamic processes. Briefly, Elga's argument begins by considering an anti-thermodynamic process. According to a standard line of reasoning in statistical physics, such processes are overwhelmingly improbable, since they are extremely sensitive to variations in microscopic *initial* conditions. Only an extremely small fraction of possible trajectories in the system's microscopic phase space correspond to anti-thermodynamic behavior, and these trajectories in general have the feature that only very slight changes in the values of the state variables will result in a trajectory corresponding to a thermodynamically normal macro process. But if the dynamics governing the system is time-reversal-invariant, this means that only slight changes in the *final* conditions of a thermodynamically normal system should result in a micro trajectory corresponding to a system with thermodynamically abnormal past (since a thermodynamically normal system is thermodynamically abnormal, if looked at backward in time). Thus, Elga, concludes, there is no asymmetry of miracles, since a small miracle is enough for two worlds with radically different pasts—a world in which the egg was cracked and one in which it was not—to converge.

How are the various traces that the cracking of the egg leaves in our world manufactured in the miracle world? If the miracle is confined to the pan, how, to take one of Elga's own examples, is an image of the egg cracking formed on a video camera of a spying neighbor? Elga does not discuss this in any detail, but our discussion in the previous section can help us understand how the image is formed. In the world where Gretta does not crack the egg, her body and the kitchen walls would—instead of acting as partial absorbers—emit coherent radiation, such that this radiation destructively interferes with the light reflected from Gretta's empty hands and mimics the light that would have been reflected by the cracking egg in the actual world.

Elga's discussion ends with the conclusion that perfect convergence is easy. Now what does this show? In particular, do the thermodynamic considerations go beyond the general argument against Lewis's thesis that I presented above, or do they simply work out the general point for a special case? Even though Elga himself does not discuss these, the thermodynamic case does suggest a number of conclusions special to it.

A worry about Elga's argument is that it seems to be showing too much. For the future of a thermodynamically normal system is not similarly sensitive to slight changes in its initial conditions as its past is to changes in final conditions. Almost all possible micro trajectories corresponding to a given macro state are thermodynamically normal in that they will not result in decreases in entropy in the near future, and nearby trajectories are likely to result in similar macro behavior. Thus, if Elga's argument is right, then the future of a system appears to be *less* sensitive to small miracle changes in initial condition than the past is. There may be certain, special, 'microscopic' miracle changes that lead to diverging macro futures. But almost any micro miracle will lead to rather strikingly diverging macro pasts. Statistical physics *alone* would suggest that it is overwhelmingly likely that any present low entropy state is a local entropy minimum—that is, is a state for which entropy increases both toward the past and toward the future. Of course in the actual world, entropy by and large decreases toward the past. But Elga's argument shows that only slight changes in the present micro state of the system would result in a state that in fact is at a local entropy minimum. In other words, slight changes in the micro state have a drastic effect on a system's past, but will be less likely to affect the system's future in a similarly dramatic fashion. While the argument seems to show that the macroscopic history of a deterministic system is extremely sensitive to the system's microscopic final conditions, a system's macroscopic future will (by the same line of reasoning) in general be much more robust.[7] The reason for this is that the system's future in the actual world is what is statistically to be expected, while its past is not, a fact that is standardly explained by appealing to low-entropy initial conditions of the universe.

Thus, Elga's thermodynamic considerations seem to suggest that there is an asymmetry of miracles after all: Almost any small microscopic miracle will radically change a system's past, while changing a system's future is much harder. And this seems to imply further that there *is* an asymmetry of counterfactuals, but one on which backtracking counterfactuals, but not forward-looking ones, come out true. For Elga's counterfactual "If Gretta hadn't cracked the egg, then at 8:05 there wouldn't have been a cooked egg on the pan" (Elga 2001, S314) seems to come out false if one applies the statistical argument consistently. Arguably, to get to a world in which the final micro state of the cooked egg is altered ever so slightly (and in which, consequently, the egg formed out of rot), seems to require less of a miracle than to get to a world in which Gretta does not crack the egg. Even a sorcerer's apprentice can produce the former kind of miracle, while the latter requires careful design and skillful wizardry, and (depending on how physically robust the process is that leads to the cracking) may require a substantially larger miracle.

Things are worse, then, for Lewis's original account than Elga suggests. The problem with Lewis's account is not that his similarity metric leads to a symmetry between converging and diverging worlds, but that in thermodynamic worlds like ours convergence is much easier than divergence is.

Similar to the case of radiation, we can ask if both this last worry and Elga's own criticism of Lewis can be met by amending Lewis's similarity metric. One suggestion might be that all we need to do is assign a higher weight to agreement in particular macroscopic matters of fact. Diverging worlds, it might seem, are much more similar to the actual world than worlds converging to it, as far as match in particular matters of fact are concerned. Consider diverging worlds first. Lewis himself maintained that divergent worlds differ to an ever greater degree from the actual world at times farther and farther away from the time of divergence—the evolutions of the two worlds truly diverge. But it is not clear that this needs to be so, at least as far as macroscopic matters of fact are concerned. In the case of electromagnetic miracles, if the miracle is performed in a region enclosed by an absorber, then small differences in the trajectory of a single charge and the associated differences in the electromagnetic field will not 'leak' outside of the absorbing region. If there is a perfectly absorbing medium in the future of the charge, then what the trajectory of that charge is, and even whether there is a radiating charge at all, makes no difference to the field in the absorber's future. The field in the future of the absorber will be zero no matter what the fields are in its past. Thus, a small miraculous change to a particle trajectory in the laboratory will not affect the distant future of the world in significant ways. Of course, differences in the incoming field will also lead to differences in the motions of at least some of the absorber particles. But through multiple collisions these differences will quickly be spread among many degrees of freedom and will neither affect the macroscopic state of the absorber nor have macroscopic effects on other systems with which the absorber might interact in the future. Similarly, whether or not Gretta cracks an egg on Tuesday will have almost no effect on events outside of Gretta's kitchen, and she herself will soon have forgotten what she ate for dinner that day. Many small miracles will have no effect on the distant macro future of the world and, hence, there will be large agreement between the macro futures of the diverging world and the actual world.

Now contrast this with the past of converging worlds. Elga argues that the anti-thermodynamic behavior of the region in which the cracked egg appears infects larger and larger regions of the world toward the past, with the effect that it is overwhelmingly probable that a converging world began its life in a maximal entropy state. If this apparently widely held view is correct, then converging worlds are radically different macroscopically from the actual world. Some argue that this difference also amounts to a difference in laws, since they maintain that the low-entropy past of the actual world has the status of a law. But perhaps it is not even necessary to settle this issue. If, as Elga and others claim, it is overwhelmingly probable that converging worlds have originated in a high-entropy state, then these worlds differ from the actual world much more widely than diverging worlds, whether we take the low-entropy initial state of the actual world to be lawful or not. For, as we have just seen, small miracles generally have negligible effects on the future of the world. There are, of course, certain small miracles that lead to radically different macro futures—at least in our corner of the universe.[8] Yet, there are many small miracles that will not lead to a divergent macro future. By contrast, if the present line of argument is correct, *almost every* small change to the present will result in a macroscopically different past.

Thus, it might seem that, contrary to what Lewis himself suggested, the thermodynamic asymmetry can underwrite an asymmetry of counterfactuals in ways the asymmetry of wave phenomena cannot. Yet crucial to this amended Lewisian account is that it is indeed the case that the anti-thermodynamic behavior of converging worlds spreads toward the past, and I am not aware of any argument that establishes this claim. An alternative suggestion would be that the anti-thermodynamic behavior is confined to a small space-time region and, as in the case of diverging worlds generally does not 'leak out' into the distant past. According to that suggestion absorbing media will act as absorbers (in the usual time sense) except for a small region in which, for example, the walls of Gretta's kitchen emit coherent radiation to mimic the visual traces of the cracking of the egg. In the more distant past, the kitchen walls were in a relatively high entropy state, as they are in the actual world.

The argument for why the anti-thermodynamic behavior resulting from the miracle cannot merely be local is, I take it, that even if the walls in the more remote past were in a *macro* state close to that of the actual world, their *micro* state will have been different and this difference will have had effects on neighboring systems. Just as the fact that the egg is not cracked in the miracle world leads to earlier anti-thermodynamic behavior in the walls, the resulting differences in the walls past micro-state lead to anti-thermodynamic behavior still earlier in the rest of the house, and other systems close by. As in the case of the egg, any slight change in the present micro-state of a system makes it overwhelmingly probable that system had an anti-thermodynamic past. And the slightly different forces exerted by the microscopically different kitchen walls on the walls' surrounding are enough to result in anti-thermodynamic behavior there as well, or so the argument goes.

Yet what is missing to make this argument convincing is some kind of rough quantitative calculation of the relevant forces involved to ensure that the interactions between real-life, relatively isolated systems are in fact sufficiently large that microscopic changes in one system have an effect on other systems that results in an anti-thermodynamic micro-evolution. The argument supposes that the anti-thermodynamic behavior that begins (or rather ends) in Gretta's kitchen infects larger and larger regions of the universe the further back in time we go. Let us consider then, for example, the point when the anti-thermodynamic behavior spreads from Earth to the entire solar system. Is it indeed plausible, as the argument ultimately must claim, that changes in the macroscopic matter distribution on Earth would have a non-negligible effect on the past evolution of the sun? Would the forces be large enough 'to knock' the sun's complete micro-state from one that has a thermodynamically normal time-evolution to one that is thermodynamically abnormal? This may or may not be so, but before we can appeal to the amended similarity metric to rescue a Lewisian account of an asymmetry of counterfactuals, we have to have a convincing argument to show why the relevant forces are large enough such that leaving a single egg uncracked will indeed 'undo' the Big Bang and result in a world with a high entropy cosmological initial state.

In sections 3 and 4 I argued that the physics of wave phenomena cannot support a Lewisian account of the causal asymmetry. Lewis's original similarity metric does not asymmetrically distinguish between worlds diverging from the actual world and worlds converging to it. What is more, Lewis's account cannot be rescued by amending the metric to include the presence or absence of coherent waves not associated with any source as a criterion of similarity between worlds. In this section I discussed another strategy for amending Lewis's account based on thermodynamic considerations instead of the physics of wave phenomena. This strategy maintains that it follows from thermodynamic considerations that the macro histories of worlds converging to ours differ radically from that of the actual world, while the macro futures of diverging worlds do not. I suggested that this strategy still owes as an argument for its central claim. The worry, of course, is that if such an argument cannot be provided—that is, if an amended similarity metric could not be motivated independently of the context of wave or radiation phenomena—then Lewis's account cannot avoid the dilemma posed in the section 4.

7. Conclusion

I have argued that Lewis's overdetermination thesis and his thesis of an asymmetry of miracles are false. The radiation asymmetry does, however, support the claim that there is (in 'ordinary' circumstances) an asymmetry of counterfactuals, if the criteria of similarity between worlds are broadened to include certain types of qualitative agreement between worlds. But since the retardation condition, on which the asymmetry relies in the case of electrodynamics, is plausibly understood as a causal condition, the corrected and amended account cannot serve to underwrite a Lewisian counterfactual analysis of causation. Ironically, given Lewis's own emphasis of the wave asymmetry over that of thermodynamics, his account might fare better once thermodynamic considerations are brought into play. If a small miracle leads to converging worlds that have anti-thermodynamic pasts and ultimately began their lives in a high-entropy initial state, then this suggests that it might be possible to amend Lewis's similarity metric to give the correct result after all, despite the fact that there is no difference in the size of miracles required for convergence and divergence, respectively. So far, however, no one seems to have given a convincing argument for this last claim. Until such an argument is forthcoming, it is doubtful that the counterfactual and causal asymmetries can be grounded along the lines Lewis suggests. That is—for now, at least—an account taking the causal asymmetry as fundamental appears to be preferable.

9

Conclusion

My main aim in the preceding chapters has been to use classical electrodynamics as a case study to raise doubts about what I take to be a widely held view on scientific theorizing (at least as far as the physical sciences are concerned). According to this view, scientific theories ought to be identified with certain mathematical structures and with a mapping function that determines the theory's ontology. The view can be expressed either syntactically, whereby the core of a theory is identified with a mathematical formalism, or semantically, whereby the core is identified with a class of models. Central to this conception is the conviction that the interpretive framework of a theory consists of nothing but a mapping function from bits of the formalism to the world. Against this I have argued that theories may include a far richer interpretive framework than the standard conception appears to allow.

For one, I have argued that a theory's mathematical formalism can be inconsistent and yet allow us to construct highly reliable models of the phenomena in a certain domain. The Maxwell-Lorentz equations and the standard formulation of energy conservation for particle-field systems are inconsistent with the existence of discrete charged particles. Nevertheless, these equations are successfully used to model electromagnetic phenomena as involving charged particles. This is possible, and 'logical anarchy' is avoided, I have argued, since the interpretive framework of classical electrodynamics contains implicit, content-based rules guiding the selective application of the theory's basic equations. Different types of models are used to represent the different causal routes by which charged particles and electromagnetic fields interact: The Maxwell equations are used to build models representing the fields caused by a given charge and current configuration, while the Newtonian Lorentz equation of motion is used to model the motion of a charged particle due to a given external field.

My aim in the second half of this book has been to show that it is indeed legitimate, and in fact preferable, to interpret classical electrodynamics causally. In contradiction to the apparently widespread view that there is no room for a 'weighty' notion of cause in fundamental science, I have argued that there are scientifically meaningful distinctions between various causal locality principles that cannot be captured in non-causal terms; and that such causal distinctions are central among the reasons typically offered for rejecting the causally non-local and backward causal Lorentz-Dirac equation of motion.

My main focus in the book's second half has been the problem of the arrow of radiation. I have argued that the most promising explanation for the temporal asymmetry exhibited by the total fields in the presence of radiating charges appeals to a causal constraint, namely that the field component physically associated with a radiating source is fully retarded. That is, each radiating source physically contributes a diverging wave to the total field, having an effect on the total field in the future of the charges' accelerating. And this constraint can explain, I have argued, that the total field in the presence of a small number of radiating sources often can be represented as fully retarded, but not as fully advanced. Moreover, no non-causal explanation of this asymmetry seems possible. This, then, is a second way in which the traditional conception of theories is inadequate: Contrary to what that conception allows, a theory's interpretive framework can be rich enough to include substantive causal assumptions.

That scientific theorizing does not merely consist in presenting a mathematical formalism delineating the possible worlds allowed by a theory has become a prominent theme in philosophical investigations of theory application and 'low-level' model-building. I have tried to show how some of the theoretical 'tools' critical to theory application—such as appeals to causal constraints, considerations of simplicity and mathematical tractability, and rules guiding the selective use of different bits of a formalism—play an important role even at the level of fundamental theory.

Notes

Chapter 1

1. See Suppe (1977) for a detailed discussion of the syntactic or 'received' view (as it used to be called) and its problems.

2. The main advocates of the semantic view in the United States are Patrick Suppes (1961, 1967, 1969, 1970), Bas van Fraassen (1970, 1972, 1980, 1989), Frederick Suppe (1989), and Ronald Giere (1988, 1999a, 1999b). A closely related view is the German *structuralism* developed by Wolfgang Stegmüller and others under the influence of Suppes's approach. See Balzer, Moulines, and Sneed (1987) for a development of the latter view.

3. The two notions are conflated, for example, in Suppes (1961), van Fraassen (1980), Giere (1988), and Lloyd (1994). For detailed textual evidence for this charge, see Frisch (1998).

4. In fact, Friedman (1982) argues that the two views are equivalent.

5. Another recent endorsement of this view is in Ismael (1999).

6. For similar lists of criteria of theory assessment, see Newton-Smith (1981, 226–232) and Darden (1991, 257–269).

7. See also Longino (2002a), Dupré (1993), and the essays in Morgan and Morrison (1999).

8. Giere now seems to disavow his earlier appeal to a model-theoretic conception of model in characterizing the semantic view; see Giere (1999a, 1999b).

9. For a critical discussion of this claim, see Sklar (2000).

10. See also Kitcher (2001) and Longino (2002a, 2002b, 2002c).

11. Kitcher's and Longino's debate appears to center largely on the question of whether different representations constructed with the help of different theories are jointly consistent—that is, the question of whether different theories need to be consistent with one another—rather than the issue of whether individual theories ought to be *internally* consistent. Depending on whether we think that there may or may not be reasonably sharp criteria that distinguish one theoretical framework from another, we may or may not think that there is a sharp distinction between the two issues. However, a case of an inconsistency clearly within what by anyone's lights would count as a single theoretical framework would be even more problematic for a "moderate methodological conservatism," such as Kitcher's, than inconsistencies between largely noninteracting theoretical schemes.

Chapter 2

A shorter version of this chapter appears as Frisch (2004-a), copyright by the Philosophy of Science Association.

1. As I said above, by dividing the discussion up in this manner, I hope to make the material accessible to a wider audience. Some readers may want to skip the detailed discussion of alternative equations of motion in the next chapter and simply accept the conditional claim: If there is no consistent classical theory to which the Maxwell–Lorentz scheme is an approximation, then the lessons I draw in this chapter from the fact that this scheme is inconsistent, are warranted.

2. Excellent textbooks on classical electrodynamics include advanced undergraduate texts such as Griffiths (1999) and (Panofsky, Hermann, and Phillips (1962), and graduate- or research-level presentations such as Landau and Lifshitz (1951) and Jackson (1975, 1999). Jackson's texts, in particular, seem to be widely regarded as the definitive treatments of the subject.

3. See Lange (2002, 112–120) for a nice, simple argument for these claims.

4. This does not mean that we all have to become scientific realists; we need not endorse the existence claims of a theory we accept.

5. See Brown (2002) for a condition like this.

6. For a recent collection of both formal and informal approaches to inconsisteny in science, see Meheus (2002).

7. See, however, Malament (1995).

8. See also the papers by Arthur Miller and Nancy Nersessian in Meheus (2002).

Chapter 3

Section 2.4 of this chapter appears in Frisch (2004-b), copyright by the Philosophy of Science Association.

1. In personal correspondence Jackson told me that his reference to the rate of work on a charge should be reworked to concern infinitesimal charges instead, and that the derivation in section 6.8 was not meant to cover compact localizations of charges. The question remains, however, why the discussion occurs in a section which is advertised as being concerned with charged particles.

2. Parrott (1987) argues for this point of view explicitly in connection with the Lorentz–Dirac theory, to be discussed below.

3. This is a consequence of the discussion in Parrott (1987, sec. 5.7). One can define an energy–momentum tensor that is consistent with the Lorentz force equation of motion by subtracting a suitably chosen acceleration-dependent, divergence-free term from the standard tensor.

4. See Sklar (1990) for an argument to that effect based on different case studies.

5. I discuss Lorentz's philosophical views in more detail in Frisch (forthcoming).

6. It appears that the stability question has been settled only relatively recently. Pearle (1982) cites and endorses Lorentz's argument. But Yaghjian (1992) showed that one can generalize Poincaré's stresses in a way that renders a relativistically rigid electron undergoing arbitrary motion stable.

7. See chapter 4 for further discussions of the role of analyticity conditions in theorizing.

8. Rohrlich (1990) shows that we get the correct limit if we ignore the origin of the total mass m in the theory and assume that the total mass is not affected by taking the neutral-particle limit. But this assumption appears to be unwarranted. If the total mass

consists of both an electromagnetic and a nonelectromagnetic component, then the former, but not the latter, should be affected by taking the neutral-particle limit. To be sure, Rohrlich claims that his own derivation of the equation of motion of a point charge does not rely on a renormalization. A central assumption in the derivation is the asymptotic condition that the acceleration of a charge vanishes at both past and future infinity (see chapter 4 for a more detailed discussion of this condition). From this, Rohrlich concludes that the momenta associated with the Coulomb fields of the charge at past and future infinity cancel (which would mean that the infinities in the field cancel without the need of invoking a negative bare mass). But this is a mistake. Since the momenta depend on the *velocity* of the charge, the incoming and outgoing momenta will in general be different, even if the *acceleration* in the infinite past and the infinite future vanishes. Thus, Rohrlich's derivation, as well, cannot succeed without invoking a mass renormalization.

9. My discussion here follows Levine, Moniz, and Sharp (1977).

10. See, for example Rohrlich (1990). Dirac calls the fact that one has to postulate a vanishing final acceleration at infinity in addition to the standard Newtonian initial conditions "the most beautiful feature of the theory" (Dirac 1938, 157).

11. Rohrlich (1990, 186) argues for this view.

12. If the force function is nonzero somewhere and analytic everywhere, then the force cannot be identically equal to zero in any interval that is a subset of the charge's trajectory. In particular, the force can tend toward zero in the infinite past, but cannot be identically zero before any finite time t. Thus, the charge cannot begin to accelerate before the onset of any external force.

13. See chapter 4 for a more detailed investigation of the causal structure of the Lorentz–Dirac theory and its connections to the analyticity requirement.

14. See also Spohn (2000). The earliest version of this type of argument of which I am aware appears in Landau and Lifshitz (1951).

15. See Rohrlich (2001). As in his discussion of the same issue in the context of the Lorentz–Dirac equation (Rohrlich 1990), Rohrlich makes the mistake of asking only what the acceleration is of a free particle that *never* experiences any force.

Chapter 4

This chapter is a much expanded version of Frisch (2002), copyright by the British Society for the Philosophy of Science.

1. One may object to my appeal to this theory in arguing for any substantive philosophical conclusions. As we have seen in chapter 3, the theory is deeply problematic and is rejected by most physicists. But part of the reason *why* the theory is rejected concerns the causal structure that is commonly attributed to the theory. Thus, it is a legitimate question to ask what that causal structure is.

2. "Characteristisch für diese physikalischen Dinge ist ferner, dass sie in ein raumzeitliches Kontinuum eingeordnet gedacht sind. Wesentlich für diese Einordnung der in der Physik eingeführten Dinge erscheint ferner, dass zu einer bestimmten Zeit dies Dinge 'in verschiedenen Teilen des Raums liegen.' Ohne die Annahme einer solchen Unabhängigkeit der Existenz (des 'So-Seins') der räumlichen distanten Dinge voneinander, die zunächst dem Alltagsdenken entstammt, wäre physikalisches Denken in dem uns geläufigen Sinne nicht möglich. Man sieht ohne solche saubere Sonderung auch nicht, wie physikalische Gesetze formuliert und geprüft werden könnten. Die Feldtheorie hat dieses Prinzip zum Extrem durchgeführ, indem sie die ihr zugrunde gelegten voneinander unabhängig existierenden elementaren Dinge sowie die für sie postulierten Elementargesetze in den unendlich-kleinen Raum-Elementen (vierdimensional) lokalisiert.

"Für die relative Unabhängigkeit räumlich distanter Dinge (A und B) ist die Idee characteristisch: äussere Beeinflussung von A hat keinen *unmittelbaren* Einfluss auf B; dies ist als 'Prinzip der Nahewirkung' bekannt, das nur in der Feld-Theorie konsequent angewendet ist. Völlige Aufhebung dieses Grundsatzes würde die Idee von der Existenz (quasi-) abgeschlossener Systeme und damit die Aufstellung empirisch prüfbarer Gesetze in dem uns geläufigen Sinne unmöglich machen." My translation follows closely that in Howard (1989, 233–234).

3. This ambiguity exists in the original German as well.

4. Lange defines spatial locality as follows: "For any event E and for any finite distance $\delta > 0$, no matter how small, there is a complete set of causes of E such that for each event C in this set, there is a location at which it occurs that is separated by a distance no greater than δ from a location at which E occurs" (Lange 2002, 14). Temporal locality is defined analogously. As Lange points out, spatiotemporal locality is logically stronger than the conjunction of spatial locality and temporal locality. What spatiotemporal locality excludes, is the (perhaps rather abstruse) possibility, compatible with the conjunction of spatial locality and temporal locality that there are causes arbitrarily close to E, but that such close causes occur a long time before E; and that there are *other* causes occurring an arbitrarily short time before E, but which occur far away from the place where E occurs.

5. Healey introduces a condition of relativistic locality as follows: "There is no direct causal connection between spacelike separated events" (1997, 25). By contrast, the condition I am interested in excludes mediated spacelike causation as well.

6. Healey, by contrast, argues that one can have local action even if separability fails (see Healey 1997).

7. Lange (2002) questions whether there are any independent arguments for demanding spatiotemporal locality, while Mundy (1989) argues that there is no sufficient scientific support for accepting energy–momentum conservation in general.

8. I owe this point to Bob Batterman.

9. See Wu and Yang (1975) and Healey (1997).

10. According to Stokes's theorem, the line integral along the closed curve is equivalent to a surface integral over the curl of A on an open surface bounded by the curve. Since the curl of a gradient vanishes, the Dirac phase factor is gauge-invariant.

11. In fact, whether or not the theory with realistically interpreted potentials is local depends on the choice of gauge. In the *Coulomb gauge*, in which the vector potential A is divergence-free—that is, satisfies $divA = 0$—the scalar potential Φ is the instantaneous Coulomb potential due to the charge density ρ. Thus, in the Coulomb gauge, Belot's criterion seems to suggest that the theory is diachronically nonlocal, since in order to find out how the scalar potential will change here, we have to 'look at' the charge density arbitrarily far from here. (The vector potential, however, is diachronically local even in the Coulomb gauge.) But in the *Lorenz gauge*, which treats scalar and vector potentials on an equal footing (and, thus, is the natural gauge to use in a relativistic setting, where both potentials are components of a single four-vector potential A^μ), both potentials satisfy a wave equation and propagate at a finite speed c. Thus, in the Lorenz gauge the theory is diachronically local.

12. See Rohrlich (1990, eqn. 6–57). The equation is in standard four-vector notation. In my presentation of the formalism and its interpretation, I largely follow Rohrlich's classic text. For another excellent discussion of Dirac's theory, see Parrott (1987).

13. Since position enters implicitly through the fields, the equation cannot be understood as a second-order equation for the four-velocity v^μ.

14. This second assumption is often not made explicitly in the literature but is clearly needed.

15. Recall that when I speak of Newton's laws, I intend this to include their relativistic generalization and do not mean to draw a contrast between nonrelativistic and relativistic physics.

16. But Rohrlich claims incorrectly that it follows from the fact that charges which never experience a force move with constant velocity that Dirac's theory satisfies Newton's principle of inertia. As one can easily see from (4.4), the acceleration of a charge in Dirac's theory can be nonzero even at times when the force on the charge is zero. Thus, Dirac's charges violate Newton's principle of inertia and satisfy only the weaker principle that the acceleration vanishes asymptotically far away from any force.

17. See Rohrlich (1990, eqn. 6–84). ξ is a number of order 1.

18. See chapter 7 for a discussion of this notion.

19. Even if Grünbaum were able to show that Diracian velocities play the role of Newtonian accelerations, this would not be enough to show that Dirac's theory is not retrocausal. For accelerations in Dirac's theory also can *change* prior to any external force. Thus, if Diracian forces caused (appropriately time-weighted) changes in the acceleration, just as Newtonian forces cause changes in velocity, the theory would still be retrocausal.

20. Thus, ironically, Grünbaum, who is accusing advocates of a backward-causal interpretation of Dirac's theory of confusing Newtonian physics with Aristotelian physics, is himself guilty of that charge.

21. See also Healey (1997, 25).

22. I discuss the notion of possible interventions in more detail in chapter 7.

23. Since one could in principle measure the acceleration by determining the radiation field of the charge, there is a rather straightforward sense in which the acceleration *here* can be looked at. Presumably everything that can be measured can be looked at, but the converse does not hold.

24. Could it not be that it is impossible for us to know what the local state of a system is, and that therefore we are unable to predict what will happen *here* next, even though the state is determined locally? But Belot's criterion is clearly not meant to be an epistemological condition, and his talk of looking and predicting must be meant metaphorically. In the way Belot uses the terms, I take it, every real property of a system can be looked at and can be used to predict what will happen next.

25. Earman himself says that the condition "captures a good part of the content of the action-by-contact principle" (Earman 1987, 455). See also the definition of a local field theory in Earman and Norton (1987).

Chapter 5

1. See Jackson (1975, sec. 6.6), Davies (1974, ch. 5), and Zeh (1989, sec. 2.1) for discussions of the relevant mathematical formalism.

2. Albert (2000) announces the fact that the Maxwell equations are not invariant under the transformation $t \to -t$ with much fanfare: "[Classical electrodynamics] is *not* invariant under time reversal. Period.... And everything everybody has always said to the contrary is wrong" (14). And a bit later: "And so (notwithstanding what all the books say) there have been dynamical distinctions between past and future written into the fundamental laws of physics for a century and a half now" (21). But contrary to Albert's suggestion, "the books" are well aware of this fact. See the quote from Zeh in the text and the discussion in Davies (1974, 24–26).

3. Recall the discussion in chapter 2 of the standard practice in electrodynamics of treating charges either as sources of fields or as being affected by the field, but not both.

4. If both F_1 and F_2 are solutions to an inhomogeneous differential equation, then $(F_1 - F_2)$ is a solution to the corresponding homogeneous equation. F_1 can be expressed as $F_2 + (F_1 - F_2)$, that is, as the sum of an arbitrary solution to the inhomogeneous equation and a solution to the homogeneous equation.

5. One might think that symmetry could be restored simply by also demanding that $A_{out}^\mu = 0$ as well. But both conditions cannot in general be satisfied simultaneously. For it would then have to be the case that $A_{ret}^\mu = A_{adv}^\mu$, which does not hold in general. (One can see this, for example, by considering a universe with a single charge.)

6. Unfortunately, I did not make this point carefully enough in Frisch (2000).

7. This worry is discussed in North (2003).

8. "Stellt man sich auf diesen Standpunkt, so nötigt die Erfahrung dazu, die Darstellung mit Hilfe der retardierten Potentiale als die einzig mögliche zu betrachten, falls man der Ansicht zuneigt, dass die Tatsache der Nichtumkehrbarkeit der Strahlungsvorgänge bereits in den Grundgesetzen ihren Ausdruck zu finden habe. Ritz betrachtet die Einschränkung auf die Form der retardierten Potentiale als eine Wurzel des zweiten Hauptsatzes, während Einstein glaubt, daß die Nichtumkehrbarkeit ausschließlich auf Wahrscheinlichkeitsgründen beruhe."

9. "Diejenigen Formen (der elektromagnetischen Gesetzmäßigkeiten) in denen retardierte Funktionen vorkommen, sehe ich im Gegensatz zu Herrn Ritz, nur als mathematische Hilfsformen an."

10. Thus, as I mentioned in chapters 2 and 4, considerations of energy conservation appear to have played a crucial role in interpreting fields realistically.

11. "Zunächst können wir, wenn wir bei der Erfahrung bleiben wollen, nicht vom Unendlichen reden, sondern nur von Räumen, die außerhalb des betrachteten Raums liegen. Ferner aber kann aus der Nichtumkehrbarkeit eines derartigen Vorgangs eine Nichtumkehrbarkeit der elektromagnetischen Elementarvorgänge ebenso wenig geschlossen werden, als eine Nichtumkehrbarkeit der elementaren Bewegungsvorgänge der Atome aus dem zweiten Hauptsatz der Thermodynamik geschlossen werden darf."

12. "Die Grundeigenschaft der Undulationstheorie, welche diese Schwierigkeiten mit sich bringt, scheint mir im folgenden zu liegen. Während in der kinetischen Molekulartheorie zu jedem Vorgang, bei welchem nur wenige Elementarteilchen beteiligt sind, z. B. zu jedem molekularem Zusammenstoß, der inverse Vorgang existiert, ist dies nach der Undulationstheorie bei elementaren Strahlungsvorgängen nicht der Fall. Ein oszillierendes Ion erzeugt nach der uns geläufigen Theorie eine nach außen sich fortpflanzende Kugelwelle. Der umgekehrte Prozeß existiert als Elementarprozeß nicht. Die nach innen fortschreitende Kugelwelle ist nämlich zwar mathematisch möglich; aber es bedarf zu deren angenäherter Realisierung einer ungeheuren Menge von emittierenden Elementargebilden. Dem Elementarprozeß der Lichtemission kommt also der Charakter der Umkehrbarkeit nicht zu. Hierin trifft, glaube ich, die Undulationstheorie nicht das Richtige."

13. In a genuine field theory that takes fields to be part of the ontology, the worry about energy conservation disappears.

14. As is well known, this was also Hempel's view. See Hempel (1965).

15. Arntzenius (1994) suggests the same reading of Popper as Price and makes the same criticism.

Chapter 6

I first discussed the puzzle of the arrow of radiation in Frisch (2000). My views have evolved significantly since then. Nevertheless, there are a few passages from that paper—in particular my criticism of Price's account of the asymmetry—that I am reprinting here with the permission of the British Society for the Philosophy of Science.

1. As we have seen in chapter 5, this interpretation of Einstein's views is rather problematic.

2. A similar criticism is made by Arntzenius (1994).

3. See the discussion of Hogarth's theory in Gold and Bondi (1967).

4. The factor of 2 is due to the fact that Hogarth is interested in the physical contribution of each charge to the field, which in the Wheeler–Feynman framework is given by $1/2F_{ret} + 1/2F_{adv}$, while the Maxwell equations tell us that if the outgoing fields are zero, the fields are fully advanced. But this is only a notational issue.

5. Price has objected to this argument by claiming that it relies on the "ad hoc and implausible" assumption that the total field acting on a test charge has two distinct components, one associated with the absorber and one associated with the source (Price 1996, 69). But the fact that the field acting on a test charge is composed of distinct components follows immediately from the assumption that the field physically associated with each charge is half retarded and half advanced. To determine the total field acting on any charge, we need to add up the advanced and retarded contributions of all the charges in the problem. And the assumption of a symmetric interaction is justified through the same kinds of considerations as any scientific theory is: The theory is motivated partly by general conceptual considerations, such as that it is symmetric just as the Maxwell equations are, and partly by appeal to its empirical adequacy in that it gives the right answer for the fields acting on a charge (see also Leeds 1994, 1995; Price 1994).

6. See Price (1996) for this objection.

7. One way in which one could try to avoid the conclusion that a thermodynamically asymmetric absorber is physically incompatible with symmetric field forces is by trying to construct a model of the absorber in which all entering fields are fully absorbed within a finite region. The upper limit of the radial integral could then be some finite radius R_{max} (instead of infinity), and the integral would be finite. A problem with this rescue attempt, however, is that Wheeler and Feynman's model of an absorber characterized by a complex refractive index would have to be given up, because that model implies that fields are damped exponentially and hence are never strictly zero for any finite radius R. Moreover, it is doubtful that any model with some finite limit of integration would readily yield exactly the right expression for the absorber response force.

8. This section closely follows the argument against Price in section 3 of Frisch (2000).

9. Claims (iii), (iv), and (v) are the claims Price (1996, 60–61), distinguishes; he numbers them (3.1), (3.2), and (3.3).

Chapter 7

1. See also Zeh (1989).

2. In his criticism of Davies, Price makes a related point (Price 1996, 55–56). Ridderbos's defense of Davies (and Wheeler and Feynman) against Price's criticism (Ridderbos 1997) is not successful, and is itself guilty of the temporal double standard Price warns against.

3. Landau and Lifshitz (1951) motivate the retarded field representation through considerations such as these.

4. S is a Cauchy surface if its domain of dependence is the entirety of Minkowski space-time. The future domain of dependence $D^+(R)$ of a subset R of space-time is "the collection of points x such that every future-directed causal curve which passes through x and which has no past end meets R." The past domain of dependence $D^-(R)$ is defined analogously (see Earman 1986, 58).

5. Thus, Earman's claim that the Maxwell equations plus the Lorentz force law "admit a well-posed initial value problem as long as the charges move with subluminal velocities" (Earman 1986, 48) is rather misleading.

6. For a more precise criterion see, for example, Woodward (2003).

Chapter 8

1. See Lewis (2000) and the other papers in the same issue of the *Journal of Philosophy* devoted to Lewis's account of causation.

2. There are two additional criteria which not need concern us here.

3. More on this point below.

4. Arntzenius briefly mentions this case in a footnote. He correctly says that radiative phenomena do not exhibit an asymmetry of overdetermination, but he also partly misdescribes the case.

5. Arntzenius (1990) makes this point as well.

6. Price makes a similar point for the case of miracles affecting the positions of particles falling in a gravitational field, and argues that in this case, convergence to a world like ours is as easy as divergence from it (Price 1992, 509–510).

7. Of course there are chaotic systems for which microscopic changes result in radically different future evolutions. The point I am making here is only that there are *also* many nonchaotic systems whose future evolutions are insensitive to microscopic changes. If Elga is right, then the past evolutions of any such system are not insensitive to micro changes.

8. Even the famous butterfly flapping its wings resulting in a storm far away has a negligible effect on the distant future of the universe as a whole.

Bibliography

Abraham, Max. 1908. Elektromagnetische Theorie der Strahlung. 2nd ed. Vol. 2, Theorie der Elektrizität. Leipzig and Berlin: Teubner.
Albert, David Z. 2000. Time and Chance. Cambridge, Mass.: Harvard University Press.
Anderson, James L. 1992. Why We Use Retarded Potentials. American Journal of Physics 60 (5): 465–467.
Arntzenius, Frank. 1990. Physics and Common Causes. Synthèse 82: 77–96.
———. 1994. The Classical Failure to Account for Electromagnetic Arrows of Time. In Scientific Failure, edited by Tamara Horowitz and Allen I. Janis, Lanham, Md.: Rowman & Littlefield.
Balzer, Wolfgang, C. Ulises Moulines, and Joseph D. Sneed. 1987. An Architectonic for Science: The Structuralist Program. Dordrecht, Netherlands: Reidel.
Belot, Gordon. 1998. Understanding Electromagnetism. British Journal for the Philosophy of Science 49: 531–555.
Bohr, Niels. 1948. On the Notions of Causality and Complementarity. Dialectica 2: 312–319.
Brown, Bryson. 2002. Approximate Truth. In Inconsistency in Science, edited by Joke Meheus. Dordrecht, Netherlands: Kluwer.
Callender, Craig. 2001. Thermodynamic Asymmetry in Time. In The Stanford Encyclopedia of Philosophy, edited by E. N. Zalta.
Cartwright, Nancy. 1979. Causal Laws and Effective Strategies. Nous 13: 419–438.
———. 1983. How the Laws of Physics Lie. Oxford: Oxford University Press.
———. 1999. The Dappled World: Essays on the Perimeters of Science. Cambridge: Cambridge University Press.
Cartwright, Nancy, Towfic Shomar, and Mauricio Suarez. 1995. The Tool Box of Science. Poznan Studies in the Philosophy of the Sciences and the Humanities 44: 137–149.
Da Costa, Newton, and Steven French. 2002. Inconsistency in Science: A Partial Perspective. In Inconsistency in Science, edited by Joke Meheus. Dordrecht, Netherlands: Kluwer.
Darden, Lindley. 1991. Theory Change in Science: Strategies from Mendelian Genetics. New York: Oxford University Press.
Davies, P.C.W. 1974. The Physics of Time Asymmetry. Berkeley: University of California Press.

Davies, Paul C.W. 1975. On Recent Experiments to Detect Advanced Radiation. Journal of Physics A8 (2): 272–280.

Dirac, P.M.S. 1938. Classical Theory of Radiating Electrons. Proceedings of the Royal Society of London A167: 148–168.

Duhem, Pierre. [1914] 1962. The Aim and Structure of Physical Theory. New York: Atheneum.

Dupré, John. 1993. The Disorder of Things. Cambridge, Mass.: Harvard University Press.

Earman, John. 1976. Causation: A Matter of Life and Death. Journal of Philosophy 73 (1): 5–25.

———. 1986. A Primer of Determinism. Dordrecht, Netherlands: Reidel.

———. 1987. Locality, Nonlocality, and Action at a Distance: A Skeptical Review of Some Philosophical Dogmas. In Kelvin's Baltimore Lectures and Modern Theoretical Physics, edited by Robert Kargon and Peter Achinstein. Cambridge, Mass.: MIT Press.

Earman, John, and John D. Norton. 1987. What Price Spacetime Substantivalism? The Hole Story. British Journal for the Philosophy of Science 38: 515–525.

Einstein, Albert. 1909a. Zum gegenwärtigen Stand des Strahlungsproblems. Physikalische Zeitschrift 10: 185–193.

———. 1909b. Über die Entwicklung unserer Anschauung über das Wesen und die Konstitution der Strahlung. Physikalische Zeitschrift 10: 817–825.

———[1916], 1961. Relativity: the Special and the General Theory. New York: Crown Publishers.

———. 1948. Quanten Mechanik und Wirklichkeit. Dialectica 2: 320–324.

Elga, Adam. 2001. Statistical Mechanics and the Asymmetry of Counterfactual Dependence. Philosophy of Science 68 (Proceedings): S313–S324.

Field, Hartry. 2003. Causation in a Physical World. In Oxford Handbook of Metaphysics, edited by Michael J. Loux and D. Zimmerman. Oxford: Oxford University Press.

Flanagan, Eanna E., and Robert M. Wald. 1996. Does Backreaction Enforce the Averaged Null Energy Condition in Semiclassical Gravity? Physical Review D54: 6233–6283.

Friedman, Michael. 1982. Book Review: The Scientific Image. Journal of Philosophy 79: 274–283.

Frisch, Mathias. 1998. Theories, Models, and Explanation. Berkeley: Department of Philosophy, University of California.

———. 2000. (Dis-)Solving the Puzzle of the Arrow of Radiation. British Journal for the Philosophy of Science 51: 381–410.

———. 2002. Non-Locality in Classical Electrodynamics. British Journal for the Philosophy of Science 53: 1–19.

———. 2004a. Inconsistency in Classical Electrodynamics. Philosophy of Science 71, No. 4.

———. 2004b. Laws and Initial Conditions. Philosophy of Science 71, No. 5.

———. forthcoming. Lorentz's Cautious Realism and the Electromagnetic World Picture. Studies in the History and Philosophy of Modern Physics.

Giere, Ronald. 1988. Explaining Science. Chicago: University of Chicago Press.

———. 1999a. Science without Laws. Chicago: University of Chicago Press.

———. 1999b. Using Models to Represent Reality. In Model-Based Reasoning in Scientific Discovery, edited by Lorenzo Magnani, Nancy J. Nersessian, and Paul Thagard. New York: Kluwer Academic/Plenum.

Gold, Thomas, and Hermann Bondi, eds. 1967. The Nature of Time. Ithaca, N.Y.: Cornell University Press.

Griffiths, David J. 1999. Introduction to Electrodynamics. Upper Saddle River, N.J.: Prentice-Hall.

Grünbaum, Adolf. 1976. Is Preacceleration of Particles in Dirac's Electrodynamics a Case of Backward Causation? The Myth of Retrocausation in Classical Electrodynamics. Philosophy of Science 43: 165–201.

Grünbaum, Adolf, and Allen I. Janis. 1977. Is There Backward Causation in Classical Electrodynamics? Journal of Philosophy 74: 475–482.
Hausman, Daniel M. 1998. Causal Asymmetries. Cambridge and New York: Cambridge University Press.
Healey, Richard. 1994. Nonseparable Processes and Causal Explanation. Studies in the History and Philosophy of Modern Science 25: 337–374.
———. 1997. Nonlocality and the Aharanov–Bohm Effect. Philosophy of Science 64: 18–41.
———. 2001. On the Reality of Gauge Potentials. PhilSci Archive (http://philsci-archive.pitt.edu/archive/00000328).
Hempel, Carl G. 1965. Aspects of Scientific Explanation. In his Aspects of Scientific Explanation. New York: Free Press.
Hogarth, J.E. 1962. Cosmological Considerations of the Absorber Theory of Radiation. Proceedings of the Royal Society of London A267: 365–383.
Howard, Don. 1989. Holism, Separability and the Metaphysical Implications of the Bell Experiments. In Philosophical Consequences of Quantum Theory: Reflections on Bell's Theorem, edited by J. Cushing and E. McMullin. Notre Dame, Ind.: University of Notre Dame Press.
Hoyle, F., and J.V. Narlikar. 1995. Cosmology and Action-at-a-Distance Electrodynamics. Reviews of Modern Physics 67 (1): 113–155.
Ismael, Jenann. 1999. Science and the Phenomenal. Philosophy of Science 66: 351–369.
Jackson, John David. 1975. Classical Electrodynamics. 2nd ed. New York: Wiley.
———. 1999. Classical Electrodynamics. 3rd ed. New York: Wiley.
Kitcher, Philip. 2001. Science, Truth, and Democracy. New York: Oxford University Press.
———. 2002a. Reply to Helen Longino. Philosophy of Science 69: 569–572.
———. 2002b. The Third Way: Reflections on Helen Longino's "The Fate of Knowledge." Philosophy of Science 69: 549–559.
Kuhn, Thomas. 1977. Objectivity, Value Judgment, and Theory Choice. In his The Essential Tension. Chicago: University of Chicago Press.
Landau, L.D., and E.M. Lifshitz. 1951. The Classical Theory of Fields, translated by Morton Hamermesh. Cambridge, Mass.: Addison-Wesley.
Lange, Marc. 2002. An Introduction to the Philosophy of Physics: Locality, Fields, Energy, and Mass. Oxford: Blackwell.
Leeds, Stephen. 1994. Price on the Wheeler–Feynman Theory. British Journal for the Philosophy of Science 45: 288–294.
———. 1995. Wheeler–Feynman Again: A Reply to Price. British Journal for the Philosophy of Science 46: 381–383.
Levine, H., E.J. Moniz, and D.H. Sharp. 1977. Motion of Extended Charges in Classical Electrodynamics. American Journal of Physics 45 (1): 75–78.
Lewis, David. 1986a. Causation. In his Philosophical Papers. Oxford: Oxford University Press.
———. [1979] 1986b. Counterfactual Dependence and Time's Arrow. In his Philosophical Papers. Oxford: Oxford University Press. First published in Nous 13.
———. 1986c. Philosophical Papers. Vol. 2. Oxford: Oxford University Press.
———. 2000. Causation as Influence. Journal of Philosophy 97 (4): 182–197.
Lloyd, Elisabeth. 1994. The Structure and Confirmation of Evolutionary Theory. Princeton, N.J.: Princeton University Press.
Longino, Helen E. 2002a. The Fate of Knowledge. Princeton, N.J.: Princeton University Press.
———. 2002b. Reply to Philip Kitcher. Philosophy of Science 69: 573–577.
———. 2002c. Science and the Common Good: Thoughts on Philip Kitcher's "Science, Truth, and Democracy." Philosophy of Science 69: 560–568.

Lorentz, H.A. [1909] 1952. The Theory of Electrons. 2nd ed. New York: Dover.
Malament, David. 1995. Is Newtonian Cosmology Really Inconsistent? Philosophy of Science 62: 489–510.
Maudlin, Tim. 1994. Quantum Non-Locality and Relativity. Cambridge: Blackwell.
Meheus, Joke, ed. 2002. Inconsistency in Science. Dordrecht, Netherlands: Kluwer.
Morgan, Mary, and Margaret Morrison, eds. 1999. Models as Mediators. Cambridge: Cambridge University Press.
Morrison, Margaret. 1999. Modelling Nature: Between Physics and the Physical World. Philosophia Naturalis 54: 65–85.
Morrison, Margaret, and Mary S. Morgan. 1999. Models as Mediating Instruments. In Models as Mediators, edited by Mary S. Morgan and Margaret Morrison. Cambridge: Cambridge University Press.
Mundy, Brent. 1989. Distant Action in Classical Electrodynamics. British Journal for the Philosophy of Science 40: 39–68.
Newton-Smith, W. 1981. The Rationality of Science. Boston: Routledge and Kegan Paul.
North, Jill. 2003. Understanding the Time-Asymmetry of Radiation. Philosophy of Science 70 N.5, 1086–1097.
Norton, John D. 1987. The Logical Inconsistency of the Old Quantum Theory of Black Body Radiation. Philosophy of Science 54: 327–350.
———. 2002. A Paradox in Newtonian Cosmology II. In Inconsistency in Science, edited by Joke Meheus. Dordrecht, Netherlands: Kluwer.
Panofsky, Wolfgang, Kurt Hermann, and Melba Phillips. 1962. Classical Electricity and Magnetism. 2nd ed. Reading Mass.: Addison-Wesley.
Parrott, Stephen. 1987. Relativistic Electrodynamics and Differential Geometry. New York: Springer-Verlag.
Pearl, Judea. 2000. Causality: Models, Reasoning, and Inference. Cambridge and New York: Cambridge University Press.
Pearle, Philip. 1982. Classical Electron Models. In Electromagnetism: Paths to Research, edited by Doris Teplitz. New York: Plenum Press.
Poisson, Eric. 1999. An Introduction to the Lorentz–Dirac Equation. LANL Preprint Archive (gr-qc/9912045).
Popper, Karl. 1940. What Is Dialectic? Mind 49: 408.
———. 1956a. The Arrow of Time. Nature 177: 538.
———. 1956b. Irreversibility and Mechanics: Reply to Richard Schlegel. Nature 178: 382.
———. 1957. Reply to E. L. Hill and A. Grünbaum. Nature 179: 1297.
———. 1958. Reply to R. C. L. Bosworth. Nature 181: 402–403.
———. 1959. The Logic of Scientific Discovery. London: Hutchinson.
———. [1935] 1994. Logik der Forschung. Tübingen: Mohr.
Price, Huw. 1991a. The Asymmetry of Radiation: Reinterpreting the Wheeler–Feynman Argument. Foundations of Physics 21 (8): 959–975.
———. 1991b. Review of Denbigh, K., and Denbigh, J., Entropy in Relation to Incomplete Knowledge, and Zeh, H. D., The Physical Basis of the Direction of Time. British Journal for the Philosophy of Science 42: 111–144.
———. 1992. Agency and Causal Asymmetry. Mind 101: 501–520.
———. 1994. Reinterpreting the Wheeler–Feynman Absorber Theory: Reply to Leeds. British Journal for the Philosophy of Science 45: 1023–1028.
———. 1996. Time's Arrow and Archimedes' Point. Oxford: Oxford University Press.
Priest, Graham. 2002. Inconsistency and the Empirical Sciences. In Inconsistency in Science, edited by Joke Meheus. Dordrecht, Netherlands: Kluwer.

Quinn, Theodore C., and Robert M. Wald. 1996. An Axiomatic Approach to Electromagnetic and Gravitational Radiation: Reaction of Particles in Curved Spacetime. Physical Review D56: 3381–3394.

Ridderbos, T.M. 1997. The Wheeler–Feynman Absorber Theory: A Reinterpretation? Foundations of Physics Letters 10 (5): 473–486.

Ritz, Walter, and Albert Einstein. 1909. Zum gegenwärtigen Stand des Strahlungproblems. Physikalische Zeitschrift 10: 323–324.

Rohrlich, Fritz. 1965. Classical Charged Particles: Foundations of Their Theory. Reading, Mass.: Addison-Wesley.

———. 1988. Pluralistic Ontology and Theory Reduction in the Physical Sciences. British Journal for the Philosophy of Science 39: 295–312.

———. 1990. Classical Charged Particles. Redwood City: Calif.: Addison Wesley.

———. 1997. The Dynamics of a Charged Sphere and the Electron. American Journal of Physics 65 (11): 1051–1056.

———. 1998. The Arrow of Time in the Equations of Motion. Foundations of Physics 28: 1045–1055.

———. 1999. The Classical Self-Force. Physical Review D60: 084017, 1–5.

———. 2000. Causality and the Arrow of Classical Time. Studies in the History and Philosophy of Modern Physics 31 (1): 1–13.

———. 2001. The Correct Equation of Motion of a Classical Point Charge. Physics Letters A283: 276–278.

———. 2002. Dynamics of a Classical Quasi-Point Charge. Physics Letters A303: 307–310.

Rohrlich, Fritz, and Larry Hardin. 1983. Established Theories. Philosophy of Science 50: 603–617.

Russell, Bertrand. 1918. On the Notion of Cause. In his Mysticism and Logic and Other Essays. New York: Longmans, Green.

Schrödinger, Erwin. 1951. Science and Humanism. Cambridge: Cambridge University Press.

Shapere, Dudley. 1984a. The Character of Scientific Change. In his Reason and the Search for Knowledge. Dordrecht, Netherlands: Kluwer.

———. 1984b. Notes Toward a Post-Positivistic Interpretation of Science, Part II. In his Reason and the Search for Knowledge. Dordrecht, Netherlands: Reidel.

Sklar, Lawrence. 1990. How Free Are Initial Conditions? Proceedings of the 1990 Biennial Meeting of the Philosophy of Science Association, Vol. 2: 551–564.

———. 1993. Physics and Chance: Philosophical Issues in the Foundations of Statistical Mechanics. Cambridge and New York: Cambridge University Press.

———. 2000. Theory and Truth. Oxford: Oxford University Press.

Smith, Joel M. 1988. Inconsistency and Scientific Reasoning. Studies in the History and Philosophy of Science 19: 429–445.

Spohn, Herbert. 2000. The Critical Manifold of the Lorentz–Dirac Equation. Europhysics Letters 50 (3): 287–292.

Suppe, Frederick. 1989. The Semantic Conception of Theories and Scientific Realism. Urbana: University of Illinois Press.

———, ed. 1977. The Structure of Scientific Theories. Urbana: University of Illinois Press.

Suppes, Patrick. 1961. A Comparison of the Meaning and Uses of Models in Mathematics and the Empirical Sciences. In The Concept and the Role of the Model in Mathematics and Natural and Social Sciences, edited by Hans Freudenthal. Dordrecht, Netherlands: Reidel.

———. 1967. What Is a Scientific Theory? In Philosophy of Science Today, edited by Sidney Morgenbesser. New York: Basic Books.

———. 1969. Models of Data. In Studies in the Methodology and Foundations of Science, edited by Suppes. Dordrecht, Netherlands: Reidel.

———. 1970. Set-Theoretical Structures in Science. Unpublished manuscript.

Van Fraassen, Bas. 1970. On the Extension of Beth's Semantics of Physical Theories. Philosophy of Science 37 (3): 325–339.

———. 1972. A Formal Approach to the Philosophy of Science. In Paradigms and Paradoxes, edited by R. Colodny. Pittsburgh, Pa.: University of Pittsburgh Press.

———. 1980. The Scientific Image. Oxford: Oxford University Press.

———. 1989. Laws and Symmetry. Oxford: Oxford University Press.

———. 1993. Armstrong, Cartwright, and Earman on Laws and Symmetry. Philosophy and Phenomenological Research 53 (2): 431–444.

Wheeler, J.A., and R.P. Feynman. 1945. Interaction with the Absorber as the Mechanism of Radiation. Reviews of Modern Physics 17: 157–181.

Woodward, James. 2003. Making Things Happen: A Theory of Causal Explanation. Oxford: Oxford University Press.

Wu, T.T., and C.N. Yang. 1975. Concept of Nonintegrable Phase Factors and Global Formulation of Gauge Invariance. Physical Review D12: 3845–3857.

Yaghjian, Arthur D. 1992. Relativistic Dynamics of a Charged Sphere: Updating the Lorentz–Abraham Model. Lecture Notes in Physics series. Berlin and New York: Springer-Verlag.

Zeh, H.D. 1989. The Physical Basis of the Direction of Time. Berlin: Springer-Verlag.

———. 2001. The Physical Basis of the Direction of Time. 4th ed. Berlin and New York: Springer-Verlag.

Index

(Note: Page numbers in italics refer to a page where a term is first introduced or explained.)

Abraham, M. 56
Absorber (see also anti-absorber) 120, 152, 182
   formal definition 121, *124*, 126, 143
   ideal vs. non-ideal 126–131, 134
   physical model of 121, 123–124, 126, 130, *132–133*, 132–139, 143
   theory of radiation 20, 104
   Zeh's definition *146*
Action-at-a-distance 19, 31, 75, 93–94
   versions of classical electrodynamics 94, 111–112, 121–122
Advanced solution 30, *106*, *174*
Albert, D. 96, 197
Analyticity 58, 62, 90, 95–96
Anderson, J. 20, 104, 118–120
Anti-absorber 124, 134, 135, 138, 149
Arntzenius, F. 166, 170–173, 200, 201, 202
Asymmetry of radiation
   as causal asymmetry 114, 116, 152–164
   and cosmological asymmetry 126–132
   Anderson's account of 118–120
   associated with elementary radiation process 109
   condition RADASYM *108*, 120, 139, 145, 147, 184
   Davies's account of 134–135
   Einstein's account of 109–114
   Hogarth's account of 126–130
   Hoyle and Narlikar's account of 130–132
   Popper's account of 114–117
   Price's account of 139–142
   Rohrlich's account of 117–118
   and the thermodynamic asymmetry 110, 113–114, 116, 122, 132–139, 140, 148–152
   Wheeler and Feynman's account of 122–126, 132–134, 137–139
   Zeh's account of 146–152
   Zeh's formulation of 107
Asymptotic condition in Lorentz-Dirac theory 59, 61–62, 84–90, 99

Balzer, W. 193
Batterman, R. 196
Belot, G. 8, 32, 79, 81–83, 91, 93, 96, 197
Bennett worlds 181
Bohr, N. 78
Brown, B. 196

Index

Cartwright, N. 10–12, 14, 20
Cauchy problem 105–106
Causal constraint on radiation fields 114, 118, 157, 160, 185
Causation 14–15, 19
  asymmetry of 4, 77, 104, 116
  backward 61, 63, 71, 85–88, 92
  as determination 78, 115
  and intervention 15, 77–78, 146, 161, 162, 163
  and locality 73–100
  and a theory's interpretive framework 73, 99
Charged dust *see* charged particle, continuous distributions of
Consistency
  as criterion of theory acceptance 9, 14, 25, 67, 70
  of the theory of charged dusts 48
Counterfactuals 88, 154, 155, 158, 165–192
Criteria for theory acceptance 9, 14, 25, 67, 70

Da Costa, N. 39
Darden, L. 193
Davies, P. 121, 122, 134, 136, 137, 146, 199
Determinism 58, 159, 185
Dirac, P. A. M. 43, 59, 74, 84, 122, 123, 124, 125, 130, 166, 198
Duhem, P. 36
Dupré, J. 193

Earman, J. 8, 58, 74, 76, 91, 95, 97–98, 197, 200
Einstein, A. 20, 75–77, 78, 88, 91, 99, 103, 109–117, 120, 153–154, 157, 201, 202
Electric charge
  continuous distributions of 27, 36–38, 48–55, 67, 71
  discrete particle 27, 33, 37–38, 51, 52, 67, 69
    as extended 56–58
    as point particle 56, 60–63, 106
    'quasi'-point charge 65
    relativistically rigid 56–57
  and self field 33, 35, 49

Electromagnetic potential 28, 31–32, 80–82
Electromagnetic world picture 8, 15–16
Elementary process of radiation *113*, 111–116, 120, 154
Elga, A. 166, 187–190
Energy conservation 29, 31, 33, 35, 43, 47, 48–51, 59, 80, 87
Energy density, expression for 29
Energy flow, expression for 29
Equations of motion
  delayed differential-difference equation 57, 58, 67, 69, 117–118
  Lorentz-Dirac equation 60, 59–63, 65, 67, 70–71, 77, 84–90, 95–99, 122–123, 133, 160, 194
  Lorentz force equation 27, 30, 64, 67, 83, 87, 160
  regularized 65, 63–66, 68, 69
  Newtonian 30, 58, 85, 86, 87

Feynman, R. 20, 32, 94, 104, 110, 112, 121–126, 129–143, 145, 146, 154, 199
  as Mr. X 136
Field, H. 20
Flanagan, E. 64–65
French, S. 39
Friedman, M. 193

Gauge transformation 31, 80, 81
Giere, R. 5, 10–11, 193
Green's function 106
Grünbaum, A. 85, 86, 87, 88, 91, 197

Hausman D. 161, 166, 185–187
Healey, R. 32, 75, 76, 81, 82, 83, 93, 198, 199
Hempel, C. 198
Hogarth, J. 121, 122, 126–131, 135–138, 146, 156
Howard, D. 196
Hoyle, F. 121, 122, 126–127, 130–131, 137, 146
Hume, D. 74, 165, 166

Inconsistency 3, 7, 9, 12–15, 17–19, 132
  and consistent subsets 40–41, 46
  of Maxwell-Lorentz theory 32–35, 67
  and paraconsistent logic 38–40

of preliminary theories 40, 43, 46
and theory acceptance 9, 25
Independence condition of causes 185–187
Inertia, principle of 66, 84, 197
  weakened principle 84, 88, 99, 199
Ismael, J. 195

Jackson, J. D. 194

Kitcher, P. 12–13, 193
Kuhn, T. 9, 14, 16

Lakatos, I. 16
Lange, M. 29, 46, 76–78, 80–81, 85, 194, 196
Leeds, S. 201
Lewis, D. 21, 165–185, 187–192, 202
Lloyd, E. 193
Local Action, principle of 75, 75–78, 157
Locality 19, 30, 73
  causal conditions of 32, 74, 93, 95–96, 100
  diachronic 79, 83, 93–98
  Earman's action-by-contact principle 97–98
  of Lorentz-Dirac equation 95–96
  in Maxwell-Lorentz electrodynamics 79–83
  relativistic 76–77, 79, 91, 93–95, 97, 98, 100
  spatiotemporal 76–79, 85, 93–95, 97–98, 100
  synchronic 79
Longino, H. 12–13, 193
Lorentz force law 27, 30, 33
Lorentz, H. A. 27, 37, 56, 57

Malament, D. 194
Maudlin, T. 92
Maxwellian electrodynamics 27
Maxwell-Lorentz equations 27
  absence of global solutions 53–54
  and locality 79–83, 193
Miller, A. 196
Miracles, asymmetry of 168, 178–181, 187–189
Model vii, 5, 6, 13–16, 90
  as maps 11, 14
  as mediators 10

model-theoretic notion of 4, 6–7, 12, 19, 36–38, 52
  as representation 6, 10, 12, 37–38, 46, 90
  state space 6, 8
Morrison, M. 10–11
Moulines, C. U. 193

Narlikar, J. V. 121, 122, 126, 127, 131, 137, 146
Nersessian, N. 194
Newtonian gravitational theory, non-locality of 73, 79
Newton-Smith, W. 195
Non-locality
  as action-at-a-distance 73, 85
  causal 61, 73, 74
  of Lorentz-Dirac equation 83–85, 92–93
  as superluminal propagation 73, 92
North, J. 198
Norton, J. 26, 40–45

Overdetermination, asymmetry of 165, 168, 169, 170, 173–177

Parrott, S. viii, 53–55, 194
Pearl, J. 15, 161
Pearle, P. 43
Poincaré, H. 36, 56, 57
Point charges
Popper, K. 16, 20, 36, 103, 114–117, 169, 200
Preacceleration 61, 62, 65
Price, H. viii, 110–112, 114, 116, 121–122, 126, 139–143, 174, 198, 200, 201

Quinn, T. 64, 65

Radiation asymmetry see asymmetry of radiation
Radiation field associated with a source
  mathematically 106, 109, 114, 129, 141–142, 156–157
  physically (See also: elementary radiation process) 109, 118, 123, 129, 141–142, 145, 147–150, 152, 155–157, 162, 164
Radiation reaction 35, 47, 51–52, 55, 58, 60, 64, 68, 83, 123, 130

Reliability 26, 42, 46
Renormalization 60, 62, 84
Retardation condition 78, 146, 152–160, 165–166, 185
Retarded solution 30, 92, 106, 173–174
Retro-causation *see* Causation, backward
Ridderbos, T. M. 201
Rigid-body mechanics 93–94
Ritz, W. 103, 109–113, 116
Rohrlich, F. 16, 20, 44, 60, 62–66, 70, 85, 90, 104, 117–118, 120, 176, 196–199
Runaway behavior 60, 61, 64
Russell, B. 20, 73, 74, 75, 78, 99–100, 115

Schrödinger, E. 78, 79, 93
self-fields *see* radiation reaction
Separability, principle of 75
Shapere, D. vii, 45
Shell crossing for continuous charge distributions 53–55
Similarity metric 21, 167, 181–185, 189–192
Sklar, L. 193
Smith, J. 41–42, 44
Sommerfeld condition 107, 113
Spohn, W. 197
Stegmüller, W. 195
Suppe, F. 5, 195
Suppes, P. 195
Synchrotron radiation 33, 51–52

temporal double standard 116, 126
Theories
  and acceptance 11–12, 41
  and consistency 4, 7, 9
  and domains of validity 15–17, 42–44, 63
  established theories 16, 44
  as mathematical formalisms 3, 8, 10, 15, 35, 77, 99, 163
  model-based view of 10–11, 46
  semantic view of 5–6
  sentence view of 5, 7
  standard conception of 3, 4, 11, 19, 77, 90
Theorizing, foundational vs. pragmatic aims 68–72
Thermodynamic asymmetry 20, 104, 110, 113, 116, 122, 135, 140, 148–149, 187, 187–192

van Fraassen, B. 5–8, 10, 14, 26, 41, 195
vending machine view of theories 10

Wald, R. 64–65
Wave equation 105
Wheeler, J. 20, 32, 94, 104, 110, 112, 121–126, 129–143, 145, 146, 154, 201
Woodward, J. 161, 202

Yaghjian, A. 62, 63, 194

Zeh, H. D. 20, 104, 105, 107, 110, 112, 119, 121, 141, 145–152, 159